# Interdisciplinary Evolution Research

## Volume 4

**About the Series**

The time when only biologists studied evolution has long since passed. Accepting evolution requires us to come to terms with the fact that everything that exists must be the outcome of evolutionary processes. Today, a wide variety of academic disciplines are therefore confronted with evolutionary problems, ranging from physics and medicine, to linguistics, anthropology and sociology. Solving evolutionary problems also necessitates an inter- and transdisciplinary approach, which is why the Modern Synthesis is currently extended to include drift theory, symbiogenesis, lateral gene transfer, hybridization, epigenetics and punctuated equilibria theory.

The series Interdisciplinary Evolution Research aims to provide a scholarly platform for the growing demand to examine specific evolutionary problems from the perspectives of multiple disciplines. It does not adhere to one specific academic field, one specific school of thought, or one specific evolutionary theory. Rather, books in the series thematically analyze how a variety of evolutionary fields and evolutionary theories provide insights into specific, well-defined evolutionary problems of life and the socio-cultural domain.

Editors-in-chief of the series are Nathalie Gontier and Olga Pombo. The Series is edited from within the Applied Evolutionary Epistemology Lab of the Centre for Philosophy of Science, more information on the lab and the Centre is available at http://appeel.fc.ul.pt and http://cfcul.fc.ul.pt

More information about this series at http://www.springer.com/series/13109

Larissa Mendoza Straffon
Editor

# Cultural Phylogenetics

## Concepts and Applications in Archaeology

Springer

*Editor*
Larissa Mendoza Straffon
Centre for the Arts in Society
Leiden University
Leiden
The Netherlands

ISSN 2199-3068 ISSN 2199-3076 (electronic)
Interdisciplinary Evolution Research
ISBN 978-3-319-25926-0 ISBN 978-3-319-25928-4 (eBook)
DOI 10.1007/978-3-319-25928-4

Library of Congress Control Number: 2015960809

Springer Cham Heidelberg New York Dordrecht London


Printed on acid-free paper

Springer International Publishing AG Switzerland is part of Springer Science+Business Media (www.springer.com)

# Preface and Acknowledgments

The construction of phylogenies, or sets of related taxa descended from a common ancestor, goes back a long way in both biology and cultural studies. Tree-shaped representations of the relatedness of entities have been equally applied to organisms (e.g. the 'tree of life') as to languages (e.g. language family trees) since the nineteenth century. Advances in the study of evolutionary processes, such as the repeated detection of horizontal gene transfer and hybridization, have expanded phylogenetic models to incorporate network graphs that express reticulation. These tree and network models, then, are used to make inferences about evolutionary relationships.

Phylogenetic methods are currently being introduced as a compelling strategy to trace and reconstruct the origin, development, distribution and interrelatedness of archaeological traditions. Whereas several researchers have already started exploring this prospect successfully, the potential of cultural phylogenetics in archaeological research is yet to be fully exploited.

The central aim of this volume is, precisely, to survey and discuss the prospects and challenges of applying cultural phylogenetics in archaeology. The invited chapters introduce the key concepts and uses of phylogenetic methods and illustrate how these can be employed by archaeologists to infer, develop and test hypotheses about the processes that originate and shape archaeological traditions, such as innovation, borrowing, diffusion, convergence and loss. As an analytical tool, phylogenetic depictions (tree and network graphs) also offer the possibility of studying the relationship between cultural evolution and innovation rates, for example, by comparing traits and estimating divergence over time.

The motivation for this book originated in the activities carried out during the project 'Implementing the Extended Synthesis in Evolutionary Biology into the Sociocultural Domain'. The project was directed by Nathalie Gontier at the Applied Evolutionary Epistemology Lab (AppEEL) of the Centre for Philosophy of Science at the University of Lisbon, Portugal, and funded by the John Templeton Foundation (grant ID 36288). From the end of 2012 until the Summer of 2013, I was involved in this project as a visiting postdoctoral researcher. In November 2012, and as part of the project, Nathalie Gontier and Emanuele Serrelli organized a symposium session titled *Cultural Transmission Studies: Tree and Network Models of Micro- and*

*Macroevolution* for the Evolutionary Anthropology Society of the American Anthropological Association at their 111th annual meeting, held in San Francisco, California (http://appeel.fc.ul.pt/sub/eve/dir/aaa/aaa2012.html). I chaired the session, which included amongst its participants prominent researchers of cultural evolution, namely, Frank Kressing, Anna Marie Prentiss and Matthew J. Walsh (authors of two of this book's chapters), along with Quentin D. Atkinson, Alberto Bisin and ourselves. The aim of the session was to examine how cultural tree and network models are constructed, what kind of data they comprise, what inferences on cultural evolution they allow and how they add to theory formation in the fields of archaeology and anthropology. These topics were revisited throughout the various academic activities carried out by AppEEL in 2013, which additionally included two Interdisciplinary Evolution Schools (http://evolutionschool.fc.ul.pt) and an International Conference on Evolutionary Patterns (http://evolutionarypatterns.fc.ul.pt), where Daniel García Rivero, Carl Knappett and Matthis Krischel, also contributors to this book, were amongst the participants.

The stimulating talks and papers that were presented in these events and the interesting discussions that followed prompted us to take a careful look at the specific uses of these models across different sociocultural disciplines. Whereas in linguistics, phylogenetic methods have already become a household practice, examples of archaeological applications, we realized, consisted of monographs and research papers scattered over several edited volumes, journals and special editions on cultural evolution. However, there were almost no anthologies devoted specifically to case studies that exemplified the benefits and discussed the pitfalls of using phylogenetic methods in archaeology. I was therefore hasty and happy to accept Nathalie Gontier's invitation to put this volume together, which we believe to be an important and long-awaited contribution to the field. I would like to thank her for that opportunity and her support throughout this process.

The completion of this book, evidently, would not have been possible without the enthusiasm and effort of the authors in providing their extraordinary contributions and the kind cooperation all the referees whose judicious suggestions have enriched the contents of our papers. I am deeply grateful to each one of them for their generous collaboration. I would also like to express my appreciation to all the 2012–2013 members of AppEEL, especially Marco Pina, Marcia Belchior and Emanuele Serrelli, all those who participated in the events mentioned above and all who have somehow been involved in the realization of this volume. In particular, my gratitude goes to Becky Ackermann, Kate Bellamy, Eugenio Bortolini, Briggs Buchanan, Ann Brysbaert, Gerrit Dusseldorp, Maria Guagnin, Hieke Huistra, Thomas A. Jennings, Roberto Martínez González, Angus Mol, Felix Riede, Stepán Rückl, Juan Francisco Ruiz López, Priscilla Schoondermark, Cameron Smith, Ilya Tëmkin and Marianna Teräväinen. A special thanks goes also to our contact persons at Springer, Sabine Schwartz, Anette Lindqvist and Martina Himberger, as well as the Springer team in India for their assistance in this project. Finally, I would like to thank the John Templeton Foundation for their financial support and my host institution in the Netherlands, the Leiden University Centre for the Arts in Society.

We have striven to put together a book that will promote discussions on the impact of using modern phylogenetic methods in archaeology. The chapters that follow should help the reader to become acquainted with the key terms, concepts and debates in cultural phylogenetics and show the prospect of integrating phylogenetic methods as an archaeological research tool, while offering a critical view of the challenges that this represents.

It is our hope that this publication will become a reliable reference for archaeologists and material culture researchers interested in applying evolutionary theory and methodologies into their work.

Leiden, The Netherlands Larissa Mendoza Straffon

# Contents

# Contributors

**Inés Caridi** Institute of Calculus, School of Exact and Natural Sciences, University of Buenos Aires, National Scientific and Technical Research Council, Buenos Aires, Argentina

**Carl Knappett** Department of Art, University of Toronto, Toronto, Canada

**Frank Kressing** Philosophy and Ethics of Medicine, Institute of the History, Ulm University, Ulm, Germany

**Matthis Krischel** Department of Medical Ethics and History of Medicine, University Medical Center Göttingen, Göttingen, Germany

**Stephen J. Lycett** Department of Anthropology, University at Buffalo, SUNY, Buffalo, NY, USA

**Matt Mattes** Department of Anthropology, The University of Montana, Missoula, MT, USA

**Anna Marie Prentiss** Department of Anthropology, The University of Montana, Missoula, MT, USA

**Daniel García Rivero** Department of Prehistory and Archeology, Faculty of Geography and History, University of Seville, Seville, Spain

**Vivian Scheinsohn** Department of Anthropology, National Institute of Anthropology and Latin American Thought, National Scientific and Technical Research Council, University of Buenos Aires, Buenos Aires, Argentina

**Randall R. Skelton** Department of Anthropology, The University of Montana, Missoula, MT, USA

**Larissa Mendoza Straffon** Leiden University, Centre for the Arts in Society, Leiden, The Netherlands

**Allison Tripp** Department of Anthropology, Chaffey College, Rancho Cucamonga, CA, USA

**Matthew J. Walsh** Department of Anthropology, The University of Montana, Missoula, MT, USA

# The Applications and Challenges of Cultural Phylogenetics in Archaeology: An Introduction

**Larissa Mendoza Straffon**

**Abstract** Inferring and explaining cultural patterns and the ways in which human groups relate and interact over large spans of time or space is one of the biggest challenges for archaeologists. When dealing with either the remote past or the present, researchers struggle to learn about the conditions and mechanisms by which cultural traits originate, move, change, and disappear. The use of phylogenetic methods, originated in evolutionary biology to measure relatedness between species, can help to make significant advances toward those aims. This introduction maps the field of cultural phylogenetics, considers its potential for archaeological research, and summarizes the proposals laid out by the contributors of this book.

**Keywords** Cultural phylogenetics • Phylogenetic methods in archaeology • Cultural evolution • Cladistics and archaeology • Archaeological phylogenies

The present volume is both timely and needed. The past couple of decades have seen an accelerated increase in the number of works and discussions on the mechanisms and processes of cultural evolution and the methodological approaches to best describe and analyze them.[1] Cultural phylogenetics has successfully emerged as one of these methods (Mace et al. 2005). Yet, for many archaeologists who are not familiar with evolutionary science, it may not be completely clear what phylogenetics is and why or how it should be applied in their field of work. The aim of this book is, on the one hand, to address precisely this issue and, on the other, to offer a selection of clear examples of what phylogenetic methods can contribute to archaeological research. This introduction will briefly present the field of phylogenetics through its key concepts and will discuss its potential applications in the human sciences in general and archaeology in particular. It will also explain the

[1] See, for instance, the works listed in García Rivero, "Darwinian Archaeology and Cultural Phylogenetics", this volume; Lipo et al. 2006; Mace and Jordan 2011; Richerson and Christiansen 2013; Shennan 2009; Whiten et al. 2011.

L. Mendoza Straffon
Leiden University, Centre for the Arts in Society,
Leiden, The Netherlands
e-mail: l.mendoza@hum.leidenuniv.nl



L. Mendoza Straffon (ed.), *Cultural Phylogenetics: Concepts and Applications in Archaeology*, Interdisciplinary Evolution Research,
DOI 10.1007/978-3-319-25928-4_1

structure of the book with a summary of the chapters that compose it and, finally, it will reflect on the future of cultural phylogenetics in archaeology.

## 1 The Tree of Life Branches Out

Phylogenetics is a field of biology that studies the diversification of organisms and the evolutionary relationships among them by inferring ancestor–descendant lines, i.e., by tracing back their links to a common ancestor.[2] Basically, a phylogeny is a hypothetical reconstruction of those evolutionary relationships, built by identifying the distribution of characters among species and inferring their development. That is, a phylogeny constitutes a hypothesis of an evolutionary history.

Phylogenetic systematics, also known as cladistics, refers to the methods of phylogenetics. The aim of these methods is to infer which organisms share ancestry with others and the amount of evolutionary changes that may have occurred within lineages. In this manner, cladistics organizes taxa in groups according to relatedness (clades), by identifying shared characters among them, which have been inherited from a common ancestor (shared derived characters, or synapomorphies). The underlying assumption is that the more synapomorphies are shared by any taxa, the more closely they are assumed to be related, which implies that they share a more recent common ancestor with each other than with any other taxon included in the analysis. For example, extant mammals are constituted by two clades: Prototheria (egg-laying mammals, like the platypus and the echidna) and Theria (live-bearing mammals). Theria again branches out into two groups: marsupials and placentals. The derived character of the marsupials is the abdominal pouch, whereas the derived character that defines placental mammals is the uterine development of the fetus. Both clades, in turn, have fur, which is a primitive character shared by all mammals (Szalay 2013). The latter, however, says nothing about the relationships within the mammal group, whereas derived characters, which vary due to evolutionary change, like those related to reproductive modes, do provide information on relatedness. Inherited characters or traits, shared between taxa and their common ancestor, like the fur and mammary glands of mammals, are called homologies. Traits that evolved independently (by convergence, parallelism or reversal) and are not shared with a common ancestor are known as homoplasies; echolocation, for instance, is a homoplastic trait that arose separately in bats and cetaceans. Determining whether a character under study constitutes a homology or a homoplasy is also important to trace evolutionary relations. Phylogenetics often uses the principle of parsimony as an optimality criterion to select among competing hypotheses (Ryan 1996). Parsimony states that the most simple and efficient hypothesis should be preferred; in this case, the hypothesis that requires the smallest number of inferred evolutionary changes is more likely to better represent the ancestor–descendant relationships among taxa.

---

[2]The definitions given in this section have been reworked after Sterelny and Griffiths 2012, 197–200.

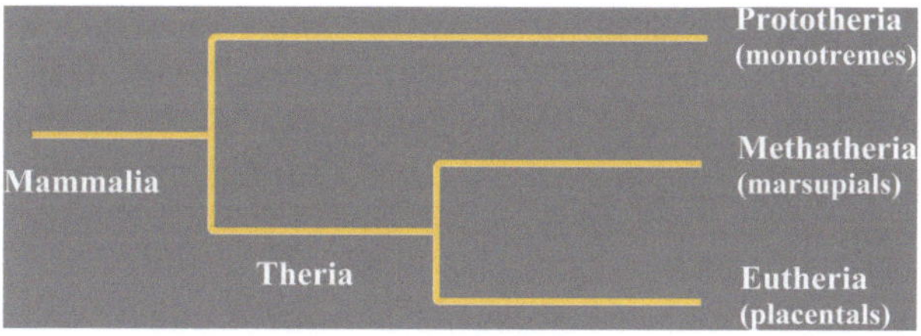


**Fig. 1** A simplified horizontal cladogram of the mammalian group (After Szalay 2013)

The result of a phylogenetic analysis is a cladogram, a branching diagram that groups taxa by shared descent. Phylogenies may be represented as trees, lines, or networks, which help visualize the processes of divergence, branching episodes, and convergence, as well as continuity or extinction. Cladograms then serve to test hypotheses about origin, relatedness, change, and, when coupled with a comparative approach, adaptation (Fig. 1). Nowadays, when dealing with large datasets of traits and taxa, most phylogenetic methods implement computational packages to generate phylogenetic diagrams.

As we have noted, the mechanisms that allow for a phylogenetic classification and analyses are common descent and variation from the ancestral form. Therefore, phylogenetic methods are applicable to any trait or entity, whether genetic or cultural, as long as it undergoes descent with modification (Levinson and Gray 2012, 167).

Like genetic information, socially transmitted information is not merely replicated from one individual or generation to the next but is also recombined and transformed, sometimes even lost. The emergence and change of cultural forms over time and space, i.e., cultural evolution, are somewhat similar to biological evolution in that it entails the basic processes of variation, selection, and transmission, i.e., descent with modification, along with others like competition, accumulation of modifications, adaptation, exaptation, and convergence (Whiten et al. 2011, 940). But biological and cultural evolution also differ in that the latter includes much higher rates of horizontal transfer, hybridization, and borrowing than its genetic counterpart, and its effects can occur much faster, in "Lamarckian" fashion (Gould 1996, 355). Furthermore, cultural evolution is not limited to continuity or extinction but can also involve the reintroduction of lost traits and reversibility to previous states (Eldredge 2000).

These differences with biological evolution have been a major point of contention in cultural evolution studies, with some scholars wondering whether cultural evolution could in fact be analogous to biological evolution and whether methods designed to study the relatively straightforward mechanisms of genetic evolution can be at all applied to analyze the various and complex modes of cultural evolution (Tëmkin and Eldredge 2007). The debates that have sprung from these issues have been widely discussed in the literature dealing with cultural evolution (Mace et al. 2005).

It is fair to say that researchers involved in cultural evolution are nowadays quite aware of these discussions and their challenges and, like the authors in this book, have taken to finding ways of integrating evolutionary methods into their studies while taking into account that these cannot be simply translated across fields but must consider the particular properties of their subject. For instance, computational and mathematical models have been adapted to specifically fit the oblique processes of cultural transmission (Whiten et al. 2011, 939). More importantly, cultural evolution researchers have been reflecting on the kinds of research questions that these methods can and cannot address (Shennan 2009).

Cultural phylogenetics, like its biological equivalent, aims at understanding cultural evolution through relations of relatedness (Currie 2013). The notion that historical social phenomena such as the origins and dispersal of populations and artifacts may be revealed by tracing back links among different cultural practices has a deep history in the human sciences, as does the arrangement of entities in sets of related taxa descended from a common ancestor (Richerson and Christiansen 2013). Tree-shaped representations of relatedness, for instance, have been common in linguistics since the nineteenth century, when the study of language family trees became a standard analytical tool in that field (see Kressing and Krischel, "Development and Degeneration: Classification and Evolution of Human Populations and Languages in the History of An-thropology", this volume). It is therefore logical that modern phylogenetic methods have been readily adopted in linguistics, especially for the purpose of classifying languages by degrees of relatedness (Levinson and Gray 2012). However, they are also being employed to study and interpret the evolution of languages, for example, by reconstructing split events between an ancestral language and daughter languages (Renfrew and Forster 2006). In this manner, several studies have constructed impressive language phylogenies that have shed light on the source and distribution of various linguistic families such as Indo-European and Austronesian language families (for a review, see Currie 2013; Renfrew and Forster 2006). More recently still, the application of phylogenetic analyses has broadened to encompass a wider array of cultural phenomena beyond languages, including social practices, and archaeological materials (Shennan 2009). In this manner, they are being used to account for human cultural variation and map out cultural histories (Whiten et al. 2011, 939).

## 2 Tracing the "Greatest Transformation" through Archaeology

The transmission and accumulation of learned knowledge and behavior across generations is what allows human culture to thrive (Tennie et al. 2009). Learning from others and the ability to share information underlie our species' success in exploiting a vast variety of environments, which has allowed it to colonize the globe in a mere hundred thousand years, or so. The quick spread and development of modern human culture that have been made possible by cumulative learning may well be "the greatest transformation in the shortest time that our planet has experienced since its crust solidified nearly four billion years ago" (Gould 1996, 354).

Those processes and mechanisms that amount to cultural evolution generate variations in behavior over time and in space (traditions) which often leave some trace in the material record (Whiten et al. 2011, 942). The analysis of that record, then, makes it possible for archaeologists to find patterns that help them reconstruct and explain the paths of cultural evolution (Mesoudi and O'Brien 2009).

Since the 1990s, several works have explored the application of these methods to the analysis of archaeological artifacts. Many of these efforts have aimed at reconstructing artifact lineages (e.g., O'Brien et al. 2001), whereas others have set out to explore the relationship between cultural evolution and innovation by interpreting artifact phylogenies to explain the mechanisms and rates of change in material culture traditions (e.g., Tehrani and Collard 2002).

The methods of phylogenetics can be put to use in archaeological research in different ways. First, like in biology, they may be used to group related artifacts or series of them (O'Brien et al. 2001). Second, they may be used to address what is known as "Galton's problem," which refers to the issue of homology vs. analogy: cultural similarities, like homologies in organisms, are unlikely to always be the result of independent invention; rather, they may be attributed to a series of other factors, such as common history, borrowing, diffusion, and coevolution. Constructing cultural phylogenies can help make inferences about those factors most likely to have shaped the traits or artifact traditions under study (Mace and Pagel 1994; Mesoudi and O'Brien 2009). Third, like in linguistics, phylogenetic methods can add to the chronological arrangement of cultural traditions (Gray and Atkinson 2003; Holden and Shennan 2005, 23). For example, by temporally situating a splitting event, a branch or node in a tree or network diagram, researchers can relatively date traits or specimens whose ages are unknown. Finally, the patterns of relatedness that emerge from such classifications and arrangements can then be used to interpret and explain archaeological cultures. For instance, they may serve as a basis to hypothesize about the distribution of material culture traditions and to test competing hypotheses (O'Brien and Lyman 2005). This last application may be the most important contribution of phylogenetic methods to archaeological research (Houkes 2011).

In recent years, several edited volumes and special issues have been devoted to discussing the theoretical, conceptual, and methodological aspects of cultural evolution in general (e.g., Mace et al. 2005; Christiansen and Richerson 2013; Shennan 2009; Whiten et al. 2011). These invariably contain papers about applications in archaeological research. However, while there are important monographs addressing the uses of phylogenetic methods in archaeology (e.g., García Rivero 2013; Mesoudi 2011; O'Brien and Lyman 2003), there are but few edited volumes dedicated to specific case studies that exemplify the potential that cultural phylogenetics holds for our field (e.g., Lipo et al. 2006). In this sense, the present book is a welcome addition.

## 3 The Contributions in This Volume

The works compiled in this book have been divided in two sections. The first part includes three topical papers (Kressing and Krischel; García Rivero; Lycett) that deal with the historical background of evolutionary thought in social science, in general, and the adoption of phylogenetic methods in archaeology, in particular.

Each, furthermore, discusses some theoretical implications of this adoption while offering an overview of the challenges and prospects of cultural phylogenetics. The second part includes a series of case studies (Knappett; Prentiss et al; Caridi and Scheinsohn; Tripp) that show how phylogenetic methods may help us ask novel questions about archaeological materials and assess hypotheses about the emergence, distribution, and transformation of archaeological traditions.

All the invited authors are outstanding and experienced researchers in the fields of cultural evolution and archaeology. The subjects they touch upon are varied, from prehistoric art to skateboards, and their geographic range is wide, spanning from Patagonia to Siberia.

The first part of the volume reflects on relevant theoretical and conceptual challenges of cultural phylogenetics. Our first contribution comprises a historical exploration of evolutionary thought in biology and anthropology. In their chapter "Development and Degeneration: Classification and Evolution of Human Populations and Languages in the History of Anthropology", Frank Kressing and Matthis Krischel examine in detail the work and legacy of several scholars who set the foundations for cultural evolution studies in the anthropological disciplines – physical and cultural anthropology, linguistics, and archaeology. They scrutinize, in particular, the theoretical dialogue between biology and social science evident, for example, in the intellectual exchange between Herbert Spencer and Charles Darwin. By focusing on two key concepts, those of development and degeneration, they provide an analytical and critical review of the history of the field since the nineteenth century. They note that development was closely linked to the idea of unilinear evolution, which was presented as an analogy of "progress" toward a higher state, whereas the horizontal transfer of traits by mixture or blending was often deemed as leading to degeneration. These ideas, the authors explain, supported a primordialist view, which purported a correlation between the purity of languages, cultures, and races. In response to this view, Franz Boas held that culture was more dependent on environment than biology, which then became the predominant paradigm in anthropology during the first three quarters of the twentieth century, relegating evolutionary thought to the background. The evolutionary revival in social science at the end of the 1970s was prompted by the new synthesis in evolutionary theory, which was consolidated in biology at that time. From that time on, trees of descent became common in physical anthropology to trace the genealogies of fossil remains and were reintroduced in linguistics to trace the origins of language families. This consequently brought about the comeback of primordialist views that attempted to plot genetic, linguistic, and archaeological lineages toward a global human phylogeny. As the authors note, the problem with this perspective is that it emphasizes the vertical transfer of languages and cultures above other forms of transmission, whereas research from cultural anthropology indicates that linguistic macro-families and genetic clusters rarely coincide and that cultural diversity predominates within those clusters. Finally, Kressing and Krischel point out that several phylogenetic studies, implicitly or explicitly, still show a primordialist undertone and challenge cultural phylogenetics to shake this view in favor of a reticulate approach. Thus, they convincingly suggest that highlighting blending and horizontal transfer would provide a more appropriate model for the actual processes of cultural transmission and development.

Continuing to lay out the historical background of the field, Daniel García Rivero, in his chapter "Darwinian Archaeology and Cultural Phylogenetics", offers a review of evolutionary thought in archaeology and of the current uses and potential of applying phylogenetic methods in archaeological research. The chapter not only looks back on previous studies but also reflects on the theoretical and methodological implications of incorporating cultural phylogenetics in archaeological analyses, in turn, responding to some of the challenges posed by the critics of this approach. The author first unravels the development of the adoption of an evolutionary paradigm in archaeology. Then, he focuses on the specific task of classifying and sequencing archaeological data to introduce the topic of cultural phylogenetics and its particular uses in studies of past material culture. In the tradition of cultural evolution theorists, García Rivero reminds us that the evolution of cultural traits minimally involves mechanisms such as variation, selection, and inheritance, and therefore, by applying Darwinian principles, it is possible to formulate and test hypotheses and explanations of cultural change. He notes, however, that much like in biology, the comparative method alone – based on formal analogy – cannot inform researchers about the nature of the similarities and differences between cultural traditions, that is, whether these are rooted in kinship, parallelism, convergence, or borrowing, what we outlined above as Galton's problem. The use of phylogenetic methods, in contrast, makes it possible to address this issue and produce sequences of classification based not only on appearance but on historical hypotheses at the population level. It is precisely this possibility of unifying phylogenetics and population thinking which holds the most promise for achieving a truly evolutionary epistemology in archaeology (Houkes 2011), and the author notes that this may constitute "the research tool of greatest potential in the field of archaeology" at this time. With a thorough overview of the existing literature, he shows that phylogenetic methods have been successfully employed in archaeology toward various aims, such as tracing cultural prototypes and their variations, identifying cultural clusters, and examining the distribution of functional traits, in order to test ideas about the temporal and geographical distribution of cultural forms and transmission mechanisms. In this manner, he concludes that the use of phylogenetics in archaeology is a fertile field that allows researchers to generate new, falsifiable historical hypotheses based on the analysis of shared derived characters to reconstruct evolutionary relations between clusters and to test them through statistical principles with quantifiable confidence levels. In sum, this chapter highlights the potential and advantages of integrating phylogenetic methods in archaeology.

In the next contribution, "The Importance of a "Quantitative Genetic" Approach to the Evolution of Artifact Morphological Traits", and following up on the theoretical and methodological reflections introduced in the previous chapter, Stephen Lycett argues for a quantitative genetic approach in evolutionary studies of material culture. He notes that archaeological studies that employ evolutionary methods, such as phylogenetics, to analyze technological relatedness, diversification, and convergence in artifactual data have been on the rise in recent years. Despite this, the theoretical development that should accompany such practices has not been advancing in equal manner. Archaeologists increasingly use analytical tools based on evolutionary principles to explore the transmission and differential persistence of material culture traditions. The underlying premise is that archaeological data can be used to infer mechanisms of social inheritance, the causes of cultural variation, and the mechanisms of selection of cultural traits, across time and space. However, archaeology still lacks a strong quantitative body of theory to link statistical variation in artifactual traits, suitable to account for different sources of cultural transmission and variation. Drawing from biology, Lycett identifies cultural information systems as the main units of cultural evolution. These are built up by socially transmitted ideas, concepts, beliefs, and practices, and their evolution involves different paths of transmission. Consequently, if the aim of our research is to reveal the evolutionary history of cultural information systems, Lycett says, we will require a quantitative framework able to shed light on "how patterns observed in physical forms over time and space respond to evolutionary forces." Quantitative genetics, he suggests, offers such framework. This approach focuses on the materiality of the data, i.e., "the physical expression of an information system that evolves via a process of descent with modification," while taking into consideration that not all variation is due to heritability. In biology, the quantitative genetic approach has brought theoretical population genetics to the study of specific physical traits, but it is also suitable for the evolutionary analysis of material artifacts. Archaeological traits are influenced by cultural and environmental effects. Quantitative genetics is able to model the total variation in an attribute present across a set of artifacts. By using a quantitative concept of heritability that allows measuring change in artifact traditions, this view can explain intergenerational variation generated by either selection or drift. Moreover, because it makes it possible to ask various questions on different sources of variation simultaneously, this view can reconcile several lines of material culture research. The advantage of the quantitative genetics approach, Lycett tells us, is twofold. On the one hand, it provides a statistical basis for studying variation in material culture, moving beyond mere "descent with modification" to reveal the statistical mechanics of changing patterns and the factors that create them. And on the other, it uses an appropriate methodology for the research object, i.e., raw archaeological data.

The second part of the book includes specific case studies and applications of phylogenetic methods in archaeology. The paper by Carl Knappett, "Resisting Innovation? Learning, Cultural Evolution, and the Potter's Wheel in the Mediterranean Bronze Age", serves to bridge between the previous historical theoretical discussions and the working examples to follow. Knappett enters a long-standing discussion in cultural evolution studies, namely, that of the nature and form of cultural transmission. He observes that most evolutionary studies of material culture have adopted a neo-Darwinian definition of learned knowledge as ideas passed from one brain to another and have consequently defined learning as a process of transmission of information. Providing examples by influential scholars to show that this is the prevailing perspective, he takes issue with the way in which the learning process and the concept of cultural trait are conceived and portrayed in much of the literature. He argues that when learning and knowledge are reduced to transmission and information, respectively, little room is left to account for the role of the environment as an active participant in the process of cultural evolution. Using the diffusion of the potter's wheel during the Aegean Bronze Age to exemplify the array of factors that intervene in cultural transmission, Knappett analyzes the possibilities of building a bridge between evolutionary and interpretive archaeologies. Based on that case study, he notes that the emergence and adoption of new technologies often depend not merely on contact between populations, as many cultural transmission studies imply, but rather rely on the local social circumstances of learning a skill, for instance, whether learning is an individual or collective enterprise in the form of an apprenticeship. In this manner, his analysis shows that the outcomes of adoption, rejection, and the rates of transmission of novel, more efficient, techniques are neither predictable nor self-evident. He therefore suggests that studies of cultural evolution should strive to encompass a broader conception of both craft knowledge and learning and proposes a shift in the explanatory focus from macro to micro level, which is better able to account for the generation, acceptance, and continuity of innovations in a local context. This, in turn, implies that by plotting actual transmission processes, researchers might end up with patterns that resemble a network rather than a tree, more suitable to accommodate lateral transfer, for example. In sum, Knappett shows that evolutionary approaches are desirable for they ask broader questions about the origins of culture and the nature of cultural change, but he makes it clear that there is also much to gain from incorporating ontogenetic processes and environmental factors into these models.

Taking up Knappett's challenge, Anna Prentiss, Matthew J. Walsh, Randall R. Skelton, and Matt Mattes, in their essay "Mosaic Evolution in Cultural Frameworks: Skateboard Decks and Projectile Points", look not only at the mechanisms of selection and drift but also at the processes of lateral transfer by cultural borrowing and independent invention. To that end, they introduce the concept of mosaic or modular evolution, which in biology refers to the process of independent change in different portions of a phenotype. The use of this concept is useful in material culture studies because, as archaeologists well know, artifacts can evolve in a modular fashion, that is, some of their components can be recombined independently. This, in fact, constitutes an important part of cumulative culture. Cultural phylogenetics, nonetheless, has not yet explored the effects of mosaic evolution, which might improve the reconstruction of cultural phylogenies. The authors reveal the effects of mosaic evolution by way of two examples. The first involves reconstructing a well-documented phylogeny. The second comprises an archaeological case study of projectile point morphology. The known phylogeny involves a cladistics and network analysis of skateboards, which confirms a branching pattern that reflects the documented history of borrowing of design components (reticulation and blending). Their archaeological example also shows that, as in the modern case, differential evolution of particular characters has probably taken place, shaping the final phylogenies. Both cases reveal that different factors such as constraints, function, and preference influence different traits in different ways, which leads to the mosaic-like selection and reproduction of characters in material culture. In addition, by means of a clever experiment using the archaeological material from a single pit house against that from other houses, Prentiss and colleagues infer aspects of vertical inheritance vs. horizontal transmission. From this, they conclude that selective pressures might be stronger not on overall morphology but on particular attributes. Some design attributes remain stable (e.g., general shape), while others present ample variation. In sum, mosaic evolution implies that human choice on variables might affect the histories of artifacts, in the past, as in the present. This requires a reassessment of the way in which cultural phylogenies for archaeological artifacts are carried out. Typically, requirements and constraints of manufacturing are grouped in single studies, whereas the present analysis suggests these had rather be looked at independently to create more accurate trees that take into account the different variables that impact cultural forms. In order to understand mosaic evolution, then, we need to understand the relationship "between modules and larger wholes and their potential for recombination over time into novel configurations." This opens the possibility of understanding divergent effects of the evolutionary forces operating on cultural forms.

The following chapter, "Mind the Network: Rock Art, Cultural Transmission and Mutual Information" by Inés Caridi and Vivian Scheinsohn offers an exciting, novel proposal to approach the effects of cultural transmission in rock art, an archaeological manifestation that has traditionally been deemed as unsuitable for these types of analyses since it lacks chronological control. These researchers, regardless, suggest that a study of motif relatedness can reveal correlations between forms allowing us to identify clusters and associations of motifs to infer patterns of cultural transmission and change. Archaeologists often use decorative patterns to sketch out cultural lineages because it is assumed that they are too complex to be recreated by mere chance, so their reproduction may be attributed to cultural transmission. In this manner, decorations have served as a basis to trace traditions of ceramics, textiles, and basketry, among others (e.g., Jordan and Shennan 2003; Shennan and Wilkinson 2001; Tehrani and Collard 2002). There is no reason why rock art motifs could not be used in the same manner, the authors argue. Using an information theory framework, they propose that each rock art motif contains information, not only in its form but also in its spatial distribution. But whereas the meaning of the earlier is hard to recover, the second may be more readily available through a correlation analysis. In this way, they establish the concept of mutual information, which, given two messages, serves as "a measure of the amount of information that one message contains about the other." This is used to describe the presence or absence of motifs across sites. Once a motif is produced in one site, it may or may not be reproduced by the same or other person in another site, soon after or at a later time, leading to the distribution pattern of a motif in the landscape. This process, by which a social network replicates a set of motifs at some point in time, is what they call a "cultural transmission path," or CTP. Mutual information allows recreating archaeological CTPs, thereby solving the problem of lacking chronological control (i.e., lack of direct dating of rock art sites and motifs). The result bears correlated sets of motifs, called mutual information networks, or MINs. Using an impressively extensive dataset of rock art for the region of Northern Patagonia, Caridi and Scheinsohn are able to successfully put together regional mutual information networks and to ultimately infer patterns of cultural transmission. They clearly show that the site network clusters obtained through their analyses cannot be simply explained by differential territorial distribution but that there is a clear correlation between certain sets of motifs. By means of these motif correlations, they are then able to determine cultural transmission archaeological paths (CTAPs) for sets of sites. Their conclusion is that the strong connectivity between regions reveals a nucleus of motifs, clustered around territories that were historically transited by Patagonian hunter-gatherers with regularity. This result is compatible with previous archaeological research, therefore lending support to a model of landscape use among local hunter-gatherer groups. This work effectively shows the potential of adopting a cultural transmission framework to ask specific archaeological questions and exemplifies several of the uses that this approach offers to archaeology, mentioned above: the possibility to group entities by means of relatedness, allowing inferences of sequence in the absence of chronologies, and testing archaeological hypotheses of cultural origin and change. In addition, it shows that rock art can be a suitable archaeological material to explore issues of cultural transmission.

Our final contribution, "A Cladistic Analysis Exploring Regional Patterning of the Anthropomorphic Figurines from the Gravettian" by Allison Tripp, deals with one of the most prominent traditions of prehistoric art, namely, the Gravettian female figurines otherwise known as Paleolithic Venuses. As she clarifies, despite the fact that these artifacts are often presented as a homogeneous set with similar stylistic attributes, there is actually almost no evidence that they indeed comprise a consistent group. In this chapter, Tripp addresses this issue by analyzing the figurines at an individual level, focusing on multiple variables to obtain a comprehensive quantitative multiregional review of the material. The aim of her study is to understand regional connections using a cladistic approach to assess two competing hypotheses about the statuettes. The first of these, suggested by André Leroi-Gourhan (1968), states that all the Venuses will share certain core features, independently of their respective geographic origin. The second hypothesis, proposed by Mariana Gvozdover (1989), establishes one stylistic group for the figurines from Russian plains, which would represent a cohesive cultural group. By determining relations of relatedness, and revealing regional patterning, a cladistics analysis can help test these propositions. Each model should lead to a different cladogram disclosing cultural evolution patterns that, in turn, can help to determine which mechanisms of transmission, vertical (branching) or horizontal (blending), had a greater influence in the Venuses' design. The underlying assumption is that higher similarity between figures from clusters of sites/regions would support horizontal transmission and blending, while higher differences between groups would support a vertical flow. Furthermore, by contrasting various formal features with an outgroup, the study is able to bring to light clear similarities and differences, on the one hand, and whether they relate to group or individual characteristics, on the other. The result of the analysis, ultimately, does not support Leroi-Gourhan's proposal (hypothesis 1) but lends some support to Gvozdover's suggestion (hypothesis 2). Thus, the phylogenetic approach reveals that the Russian figurines do seem to group together to the exclusion of other regions, indicating a regional cultural tradition. This paper provides another excellent example of how phylogenetic methods can help us test and contrast specific archaeological hypotheses, demonstrating that classical archaeological models will not always hold up to a phylogenetic analysis.

## 4 Outlook and Future Directions

The previous summary of the book's chapters discloses several common themes and offer an outlook on the challenges and directions that the field of cultural phylogenetics is taking in archaeology. Reading through the contents, it becomes clear that there are certain recurring themes permeating across studies. One, clearly, is developing new frameworks that can prevail over the shortcomings and criticisms directed at cultural phylogenetics so far. On these lines, three chapters put forward novel perspective that can potentially offer more realistic depictions of the complex processes of cultural evolution, especially in regard of the horizontal transmission of traits and traditions, and the distinction between homologies and homoplasies (Knappett; Lycett; Prentiss et al.). Another theme is exploiting the potential of the phylogenetic methods to formulate new questions about archaeological materials and test-established hypotheses that explain patterns of cultural origin, development, and loss (Caridi and Scheinsohn; Knappett; Tripp). Lastly, the more theoretically inclined papers argue for ways of achieving more accurate phylogenies and a more sophisticated evolutionary framework able to reveal and account for population-level mechanisms (García Rivero; Kressing and Krischel; Lycett).

Some of the papers presented in this book, furthermore, have embraced and answered the challenges that critics have posed to our field in the past, such as building better models of cultural transmission (e.g., network or web models) that account for all the different aspects that can shape cultural evolution, like environment, demography, and spatial relations, to mention a few. Kressing and Krischel, García Rivero, Lycett, and Prentiss et al. all have engaged with these issues. Knappett, furthermore, has offered his own critical view and invites fellow researchers to reflect on the different modes and rates of cultural change, for example, by examining the implications of adopting or rejecting a new technology vs. adopting a new artifact form or new function.

As some discussants have predicted (Houkes 2011), it seems that archaeologists will look to phylogenetics for interpreting the patterns by which cultural traits evolve, more often than they will for classification. But their success will partially depend on the accuracy of the phylogenies they use in their interpretations. Therefore, they ought to use all the information available to them to reconstruct the most factual cultural phylogenies possible. This issue has been properly explored by Knappett, Lycett, and Prentiss et al., while Caridi and Scheinsohn demonstrate precisely how researchers can draw from various sources to amass relevant data suitable for building explanations of cultural change.

The usefulness of cultural phylogenetics for testing hypotheses in archaeology should not be limited to models of common descent vs. independent evolution. As the papers by Prentiss et al., Caridi and Scheinsohn, and Tripp let see, they can also be used to test coevolutionary and convergent models, by showing how the presence of one trait or tradition influences or constraints the evolution of another and that key innovations will likely affect the rates of change and diversification of culture.

To be sure, this book does promote the implementation of phylogenetic models in archaeology. However, it also recognizes that phylogenies of material culture, like typologies and chronologies, should not become the ultimate aim of archaeological research but a regular component of the archaeologist's toolkit, one that can aid toward a better understanding of the human past.

In sum, our contributions demonstrate that the application of cultural phylogenetics in archaeology can contribute to research minimally in four ways: (1) as an aid in the classification of artifacts by relatedness of particular traits, (2) by testing hypotheses about cultural relatedness, (3) in the absence of a complete record, inform about the sequence of changes in cultural forms, and (4) providing a sound basis for interpretation and explanation of archaeological phenomena. Through these four aspects, phylogenetics provides archaeology with an important tool to reflect on how the diversity and similarity of cultural traits has evolved throughout human history.

It is foreseeable that the theoretical debates surrounding cultural evolution will continue to discuss traditional problems such as the compatibility of biological and cultural processes and the nature of the mechanisms of transmission. Within archaeology, however, it seems that the major questions will continue to be the usefulness and potential contribution of the phylogenetic methods to the field. It is our hope that this volume will become a part of those discussions, as an effective example of the successful application of the concepts and techniques of cultural phylogenetics in archaeological research.

**Acknowledgments** This introduction was written with the support of the John Templeton Foundation (grant ID 36288). I thank Ilya Tëmkin, Nathalie Gontier, and the contributors for commenting on a previous draft. I am very grateful to the series editors, Nathalie Gontier and Olga Pombo, for having trusted me with the task of editing this book and in particular to Dr. Gontier for her help throughout this project.

## References

Currie, T. E. (2013). Cultural evolution branches out: The phylogenetic approach in cross-cultural research. *Cross-Cultural Research*, *47*(2), 102–130.

Eldredge, N. (2000). Biological and material cultural evolution: Are there any true parallels?. In F. Tonneau, & N. S. Thompson (Eds.), *Perspectives in ethology* (pp. 113–153). Springer, US.

García Rivero, D. (2013). *Arqueología y Evolución: A la búsqueda de filogenias culturales*. Universidad de Sevilla, Spain.

Gould, S. J. (1996). *The mismeasure of man*. WW Norton & Company, New York.

Gray, R. D., & Atkinson, Q. D. (2003). Language-tree divergence times support the Anatolian theory of Indo-European origin. *Nature*, *426*(6965), 435–439.

Gvozdover, M. (1989). The typology of figurines from the Kostenki Paleolithic. *Anthropology and Archeology of Eurasia*, *27*(4), 32–94.
Holden, C. J., & Shennan, S. J. (2005). Introduction to part I: How tree-like is cultural evolution. In R. Mace, C. J. Holden & S. J. Shennan (Eds.), *The evolution of cultural diversity: a phylogenetic approach* (pp. 13–29). Left Coast Press, Walnut Creek, CA.
Houkes, W. N. (2011). Tales of tools and trees: phylogenetic analysis and explanation in evolutionary archaeology. In H.W. de Regt, S. Hartmann & S. Okasha (Eds.), *EPSA philosophy of science: Amsterdam 2009* (pp. 89–100). Springer, Netherlands.
Jordan, P., & Shennan, S. (2003). Cultural transmission, language, and basketry traditions amongst the California Indians. *Journal of Anthropological Archaeology, 22*, 42–74.
Leroi-Gourhan, A. (1968). *The art of prehistoric man in western Europe*. London: Thames & Hudson.
Levinson, S. C., & Gray, R. D. (2012). Tools from evolutionary biology shed new light on the diversification of languages. *Trends in Cognitive Sciences*, *16*(3), 167–173.
Lipo, C. P., O'Brien, M. J., Collard, M., & Shennan, S. J. (2006). Mapping our ancestors. In *Phylogenetic approaches in anthropology and prehistory*. London: AldineTransaction.
Mace, R., & Jordan, F. M. (2011). Macro-evolutionary studies of cultural diversity: A review of empirical studies of cultural transmission and cultural adaptation. *Philosophical Transactions of the Royal Society B: Biological Sciences*, *366*(1563), 402–411.
Mace, R., & Pagel, M. (1994) The comparative method in anthropology. *Current Anthropology*, *35*(5), 549–564.
Mace, R., Holden, C. J., Shennan, S. (2005). *The evolution of cultural diversity: A phylogenetic approach*. Left Coast Press, Walnut Creek, CA.
Mesoudi, A. (2011). *Cultural evolution: How Darwinian theory can explain human culture and synthesize the social sciences*. University of Chicago Press, Chicago, IL.
Mesoudi, A., & O'Brien, M. J. (2009). Placing archaeology within a unified science of cultural evolution. In S. Shennan (Ed.), *Pattern and process in cultural evolution* (pp. 21–32). University of California Press, Berkeley, CA.
O'Brien, M. J., & Lyman, R. L. (2003). *Cladistics and archaeology*. University of Utah Press, Salt Lake City, UT.
O'Brien, M. J., & Lyman, R. L. (2005). Cultural phylogenetic hypotheses in archaeology: Some fundamental issues. In *The evolution of cultural diversity: A phylogenetic approach* (pp. 85–108). London: UCL Press.
O'Brien, M. J., Darwent, J., & Lyman, R. L. (2001). Cladistics is useful for reconstructing archaeological phylogenies: Palaeoindian points from the southeastern United States. *Journal of Archaeological Science*, *28*(10), 1115–1136.
Renfrew, C., & Forster, P. (2006). Introduction. In P. Forster, & C. Renfrew (Eds.), *Phylogenetic methods and the prehistory of languages* (pp. 1–8). McDonald Inst of Archeological Research, Cambridge, UK.
Richerson, P. J., & Christiansen, M. H. (2013). *Cultural evolution: society, technology, language, and religion*. MIT Press, Cambridge, MA.
Ryan, M. J. (1996). Phylogenetics and behavior: some cautions and expectations. In E. P. Martins (Ed.), *Phylogenies and the comparative method in animal behavior* (pp. 1–21). Oxford University Press, Oxford, UK.
Shennan, S. (2009). *Pattern and process in cultural evolution*. University of California Press, Berkeley, CA.
Shennan, S., & Wilkinson, R. (2001). Ceramic style change and neutral evolution: A case study from Neolithic Europe. *American Antiquity, 66*(4), 577–593.
Sterelny, K., & Griffiths, P. (2012). *Sex and death: An introduction to philosophy of biology*. University of Chicago Press, Chicago, IL.

Szalay, F. S. (2013). Metatherian taxon phylogeny. Evidence and interpretation from the cranio-skeletal system. In *Mammal phylogeny: Mesozoic differentiation, multituberculates, monotremes, early therians, and marsupials* (pp. 216–242). Springer, New York, NY.
Tehrani, J., & Collard, M. (2002). Investigating cultural evolution through biological phylogenetic analyses of Turkmen textiles. *Journal of Anthropological Archaeology, 21*(4), 443–463.
Tëmkin, I., & Eldredge, N. (2007). Phylogenetics and material cultural evolution. *Current Anthropology, 48*(1), 146–154.
Tennie, C., Call, J., & Tomasello, M. (2009). Ratcheting up the ratchet: On the evolution of cumulative culture. *Philosophical Transactions of the Royal Society B: Biological Sciences, 364*(1528), 2405–2415.
Whiten, A., Hinde, R. A., Laland, K. N., & Stringer, C. B. (2011). Culture evolves. *Philosophical Transactions of the Royal Society B: Biological Sciences, 366*(1567), 938–948.

# Part I
# Concepts and Theories

# Development and Degeneration: Classification and Evolution of Human Populations and Languages in the History of Anthropology

**Frank Kressing and Matthis Krischel**

**Abstract** This contribution shows that evolutionary thought which dominated the discourse on the development of human populations, cultures, and languages in the nineteenth century (1) dates back to pre-Darwinian concepts that emerged in the times of the Enlightenment, (2) was only possible due to an ongoing interdisciplinary exchange between different branches of anthropology, and (3) was bound to the idea that lateral exchange of "racial," linguistic, or cultural traits would contribute to degeneration instead of "progressive" development. Specifically, we would like to draw the reader's attention to two quite contradictory strains in the history of science:

Evolutionary thought dominated the discourse on the development of human populations, cultures, and languages in the nineteenth century. According to this "leitmotif," inheritance took place in unilinear trees of descendence, with selection and processes of vertical descent leading to development in consecutive stages. Horizontal or lateral transfer, on the contrary, for example, of words between languages, or interbreeding between different species, populations, or "races" would ultimately lead to degeneration instead of development, spoiling the supposedly "pure" lineages of descent.

On the other hand, the development of evolutionary theory that had come to dominate scholarly thought in biology, anthropology, linguistics, and sociology could only emerge due to an ongoing interdisciplinary exchange between different branches of the sciences and the humanities, with a decisive role played by anthropology and allied disciplines. This means evolutionary theory favoring pure lines of vertical descent could only develop due to frequent and ongoing "interbreeding" between different scholarly disciplines, thus "spoiling" the pure lines of scientific descent! This interdisciplinary, "horizontal" descent is illustrated by the fact that the

---

F. Kressing (✉)
Institute of the History, Philosophy and Ethics of Medicine, Ulm University, Ulm, Germany
e-mail: frank.kressing@uni-ulm.de

M. Krischel
Department of Medical Ethics and History of Medicine, University Medical Center Göttingen, Göttingen, Germany
e-mail: Matthiskrischel@gmail.com; Matthis.krischel@medizin.uni-goettingen.de



L. Mendoza Straffon (ed.), *Cultural Phylogenetics: Concepts and Applications in Archaeology*, Interdisciplinary Evolution Research,
DOI 10.1007/978-3-319-25928-4_2

idea of biological evolution dates back to pre-Darwinian concepts that emerged in the Enlightenment and was first introduced to sociology and the humanities before being applied to the newly emerging discipline of biology in the early nineteenth century. While natural history can be traced back much further, the term "biology" was only established at that time by physicians and naturalists like Beddoes (1799), Burdach (1800), and Lamarck (1802).

This "horizontal transfer" of ideas transgressing the borders between the sciences and humanities persisted even in periods of rejection of evolutionism in both biology and cultural anthropology. We refer to "anthropology" in a broad sense, combining sociocultural anthropology with biology-derived physical anthropology and also including the neighboring disciplines of archaeology and linguistics in accord with the four field approach of North American anthropology. While the borders of "anthropology" in this sense prove to be hard to define, we understand this as just another indication for the transgression of academic borders and interdisciplinary networking between scholars – a central topic to be put forward in our paper.

**Keywords** History of Anthropology • Theory of Anthropology • Evolution • Degeneration • Primordialism • Racism • Linguistics

## 1 Introduction[1]

Theories of cultural, linguistic, and biological evolution dominated the discourse in the human sciences from the 1850s onward. In this contribution, we intend to reconstruct the development of evolutionary thought in both the physically and culturally orientated branches of anthropology and to point out mutual relations in the history of both the sciences and humanities.[2] We have identified the terms "development" and "degeneration" as central to this discourse. The two terms were by no means exclusively used in anthropology, but also in neighboring disciplines, such as evolutionary biology, comparative linguistics, and sociology. Exactly for this reason, we have structured this paper around these two terms, in order to capture the

[1] The authors would like to express their gratitude to the anonymous reviewers of this paper for providing them with very valuable advice and to the German Federal Ministry of Education and Research (BMBF) for generously funding the research project "Evolution and Classification in Biology, Linguistics, and the History of the Sciences" as part of the focus "Interaction between Natural Sciences and the Humanities" in the years 2009 to 2012. The paper presented here is based on results of this research project. Furthermore, we like to thank Jacob Tomala, student of dentistry at Ulm University, for his very insightful remarks and proofreading work.

[2] Some of the thoughts expressed in this contribution have, in a less developed manner, been expressed previously (Kressing et al. 2013), where they form the background for a critical examination of the concept of a "global phylogeny" of human populations and languages.

many-faceted influences on theories of classification and evolution of human populations and languages.

Since the scholarly discipline of anthropology is transgressing the border between the sciences and the humanities, we use the example of anthropology to point out that only mutual and perpetuated networks of scholars of both realms enabled the development of evolutionary thought as *leitmotif* of the nineteenth century (Krischel et al. 2011). Based on extensive literature research, we would like to put forward the thesis that even though physical and cultural anthropologies were institutionalized separately in the course of that century, a mutual transfer of ideas and conceptions between both realms was always maintained and that this "lateral transfer" (Fangerau et al. 2009; Fangerau et al. 2013; Krischel and Kressing 2014) between academic disciplines can be – quite conveniently – demonstrated by focusing on ideas of evolutionary development and degeneration of human populations, languages, and cultures. By lateral transfer between disciplines, we refer to interdisciplinary contacts between scholars and the exchange of ideas and methods which can, occasionally, even form "a bridge between the sciences and humanities" (Kressing 2013, p. 97).

We will further show how the idea of evolution declined in biology, linguistics, and cultural anthropology from the beginning of the twentieth century onward, achieving fresh attention with the emergence of human genetics and linguistic long-range comparison in the last quarter of the twentieth century. Again, we claim that this new focus on the idea of coevolution of human languages and genetic clusters is due to perpetuated "lateral transfer," that means mutual exchange between different academic disciplines in both the sciences and humanities (Krischel and Fangerau 2013), leading to the conclusion that anthropology as a whole always tended to include physically and culturally orientated aspect of scholarly thought from its very beginning and was always transgressing the borders of different academic disciplines.

After some remarks on classification and a short chapter on the development of evolutionary theories in both the sciences and the humanities in the eighteenth and nineteenth centuries, we will focus on the idea of co-development of language, culture, and "race" according to primordialist views (Kressing et al. 2013). By "primordialism" we refer to the conceptualization of ethnicity as "based in biology and determined by genetic and geographical factors" which is rooted in Herder's neoromantic concept of the Volk (Sokolovskii and Tishkov 1996, pp. 190–191). Since the theoretical framework of primordialist views of ethnicity is strongly influenced by evolutionism, the notion that biological and linguistic features are neither always linked nor always vertically inherited contributed to the decline of evolutionism and primordialism in anthropology (Boas 1913, 1940) and linguistics (de Saussure 1916; Trubetzkoy 1930; Jakobson 1931). We will present evidence for this decline in the early twentieth century, and the fact that the emergence of anti-evolutionism and cultural particularism in cultural anthropology and of structuralism in linguistics was accompanied by a decline of evolutionism in biology (Bowler 1983). Only after the rediscovery of Mendel's laws of inheritance, classical genetics became unified with Darwinian evolutionism in what is called the "modern evolutionary synthesis." When the neo-Darwinian paradigm had become firmly established in the 1940s, it was restricted to biology, abstaining from a primordialist connection between races and languages that had been discredited after World War II. However, primordialist ideas experienced a Renaissance in the shape of the "new synthesis" between genetic, linguistic, and archaeological data during the 1980s, leading to a model of global phylogeny. The aim of this research program was to investigate if a connection between linguistic macro-phyla and genetic clusters of humankind could be identified. Finding such a connection would lead credibility to primordialist thinking in the sense that a close link between the vertical transfer of human genes and languages would be implied. The following chapters intend to present a brief outline of the tradition of primordialist thinking in human biology, anthropology, and linguistics.

We will begin by looking at the history of anthropology as an academic discipline and shift our attention to linguistics (which was incorporated into anthropology in the early twentieth-century four field approach of U.S. American anthropology) and then to sociology. Both linguistics and sociology are scholarly disciplines that are closely related to cultural and social anthropology, as they deal with cultural features of human populations. Our basic assumption is that anthropology with its integration of a scientific, biologically orientated branch (physical anthropology) and a culturally orientated branch (cultural and social anthropology, ethnography) could only emerge as an academic discipline transgressing the science humanities borderline and that the complete "divorce" of cultural/social anthropology from physical anthropology has never been achieved (see Hann 2005, pp. vii–ix, Silvermann 2005, p. 257).

## 2 Development

### *2.1 The Development of Anthropology as an Academic Discipline*

Anthropology as "the science of man" originated in the Renaissance as an attempt of scientific emancipation from theology. The usage of the word itself (in the form "anthropologium") dates back to Hundt (1501), who defined anthropology as the science "de hominis dignitate, natura et proprietat." Almost 100 years later, Cassmann (1594) described anthropology as the lore of human nature. According to him, human nature constituted a "double form of existence, being bound to the world's spiritual as well as to the world's physical essence."[3] This can be seen as the main reason why the term anthropology, for a long time, maintained a twofolded meaning, relating to the physical as well as to the mental sphere of human existence (Streck 2000, p. 141). Even though the dichotomy between body and soul can

[3] Original quote: "Eine Wesenheit, die der doppelten Welt-Natur, der geistigen und der körperlichen, die zu einem Grundbestand vereinigt sind, teilhaftig ist."

already be found in Platonic philosophy, Cartesian dualism (Descartes 1641; Zittel 2009) renewed interest in this problem from the seventeenth century onward and advanced the split of anthropology into two distinct branches.

Physical anthropology emerged during the Enlightenment in the shape of "natural history of man" (Buffon 1749–1804), fostered by the works of Kant (1775, 1785), Blumenbach (1775), Herder (1772), Forster (1786), and Oken (1809–1811). The new subdiscipline developed three research areas: (1) the establishment of a profound image of man by comparing humans and animals, (2) medical research in human anatomy and morphology, and (3) research in the geographic variability of humans (Hoßfeld 2005, p. 57). As a science dedicated to the study of physical human nature, physical anthropology was increasingly institutionalized in the course of the nineteenth century. Subsequently, it became informed by biological theories of evolution (Darwin 1859, 1871) and eventually became part of human biology.

Around the same time, culturally orientated anthropology split from physical anthropology and focused on specifically human field of culture. It was developed as an independent scholarly discipline due to the works of Klemm (1843), Tylor (1871), Frazer (1890), and Bastian (1860). However, a complete split of the study of humankind into physical anthropology (falling into the realm of the sciences) and cultural anthropology (falling into the realm of the humanities) has never been fully accomplished. Instead, frequent transgressions of the border between physical (hominid) and cultural (humanid) anthropology can be witnessed, signified by intellectual and personal reticulation between the representatives of the two branches.

Thus, anthropology as an interdisciplinary approach to the study of human nature and culture reflects how the idea of evolutionary development took hold in both the sciences and the humanities from the Enlightenment onward. Furthermore, the example of anthropology shows that there were mutual influences of evolutionary ideas in biology, anthropology, and comparative linguistics. When we look at pre- and post-Darwinian theories of development in the eighteenth and nineteenth centuries, we notice that:

1. Development of different cultures was imagined through similar stages, i.e., savagery, barbarism, and civilization (cf. Condorcet 1795; Comte 1830–1842; Tylor 1871; Morgan 1877). In these stage models, scientific, social, and moral progress were bound to one another.
2. Cultural and linguistic development was linked in models of language typology and primordialist models of culture, examples being Herder (1772), Schlegel (1808), and Humboldt (1836).

## 2.2 *The Development of Physical Anthropology: Applying Biological Systems of Classification to Humans*

From the seventeenth century onward, Europeans had acquired a profound knowledge concerning worldwide human biodiversity (Brues 1977: 19). Concomitantly with the classification of languages, various systems of classification of the major "races" of mankind were established by comparing human phenotypes according to biometric data (Linneaus 1735; Blumenbach 1775; Kant 1775; Carus 1849; Gobineau 1853/1855). Classification schemes of the seventeenth to the nineteenth centuries considered – with varieties – three major races, e.g., Europeans or "Caucasians," Africans, and Mongoloid Asians, with different status ascribed to East Indians, Southeast Asians, and Native Americans. Among these classifications, we will focus on Blumenbach's, because he coined the term "Caucasian." Furthermore, his classification of five human "races" – Africans, Caucasians (including people from North Africa and the Middle East), Mongols, American Natives, and "Malays," e.g., Southeast Asians – resembles modern approaches to genetic sub-grouping of humans (Cavalli-Sforza et al. 1988). Blumenbach also maintained the monogenetic origin of all human populations and pointed out that there were no "pure" races, but large transitional zones of variability between them. He opposed the view that different "races" formed progressive steps of human development, including physical characteristics and intellectual capabilities. This view was clearly compatible with Enlightenment theories of cultural evolution, which had established a progressive ladder that populations would climb. By applying biological theory of evolution to humans, the development of the human species tended to be seen as part of a broader evolutionary process, explaining the emergence of different human "races" from a common ancestor. Darwin himself was a champion of monophyletic origins of humankind (Desmond and Moore 2009). Nevertheless, well into the second half of the nineteenth century, the notion prevailed that presumably "lower human races" had a shorter distance in terms of evolutionary development to primates than "higher races." Thus, racial superiority or inferiority was linked to the assumed position of a respective population in the evolutionary tree of primates and this population's distance to other, non-hominid primates.

## *2.3 The Development of Comparative Linguistics: Classifications of Human Languages*

Attempts to classify human languages and to model their development date back at least to biblical times. Until the seventeenth century, a monophyletic view of linguistic evolution was predominant. Scholars who subscribed to this view assumed that all human languages originated in Hebrew (Postel 1538; Schlözer 1781). Even though the first phylogenetic tree showing the relationship between different languages is often attributed to Schleicher (1861), a monophyletic representation by Gallet that shows the development of human languages from a common "langue primitive" can be found as early as 1800 (Arroux 1990). This is around the same time that tree models of evolutionary development start to appear in biology. Prior to treelike representations, polyphyletic models of language origin were developed from the seventeenth century onward, for example, by Boxhorn who was the first to identify the family of languages nowadays referred to as "Indo-European." In 1647, Boxhorn formulated a theory of a "family of genetically related languages deriving from a common ancestral language and distinct from other linguistic families"

(Driem 2005, p. 289, 2007, p. 211). Also in the seventeenth and eighteenth centuries, the Uralic and Altaic languages were identified as linguistic stocks different from Indo-European and Semitic (Witsen 1692; Schlözer 1781). Thus, the idea of polyphyletic origins of different languages had already been clearly expressed even before Jones (1746–1794) described the common root of Sanskrit, Greek, and Latin in 1786, an event that is often considered as the starting point of comparative studies in Indo-European languages.

In the course of the nineteenth century, historical comparative linguistics became further established, resulting in the reconstruction of protolanguages according to laws of regular sound correspondences (Bopp 1816; Rask 1818; Grimm 1819–1834). Although the relationship between most of the languages today referred to as "Indo-European" had already been well established, a debate arose concerning the name of this language family, with "Scythian," "Indo-Germanic," "Sanskritic," and "Aryan" being among the candidates (Malte-Brun et al. 1810–1826; Young 1813; Müller 1855).

Early models of morphological language typology, dating back to Smith (1762), Herder (1772), Schlegel (1808), and Humboldt (1836), emphasize the linear development from analytic, presumably "simple" built languages (e.g., Chinese) to "higher," more complex forms like the agglutinative idioms (e.g., Native American languages) and finally to inflective languages like Semitic or Indo-European. This evolutionary scheme applies cultural value to a specific type of language morphology and has therefore been severely criticized by Edward Sapir (1884–1939) a century later as a proponent of anti-evolutionism and cultural relativism:

> This is the evolutionary prejudice which instilled itself into the social sciences towards the middle of the last century and which is only now beginning to abate its tyrannical hold on our mind […]. The vast majority of linguistic theorists themselves spoke languages of a certain type, of which the most fully developed varieties were the Latin and Greek that they had learned in their childhood. It was not difficult for them to be persuaded that these familiar languages represented the 'highest' development that speech had yet attained and that all other types were but steps on the way to this beloved 'inflective' type. (Sapir 1921, pp. 132–133)

## *2.4 Primordialism: Linking "Races" and Languages: Interdisciplinary Transfer in the Service of "Pure Blood Lines"*

The basic idea to identify linguistic with physiological and cultural qualities of human populations originated in German Romanticism of the turn of the eighteenth to the nineteenth centuries. Prominent representatives included Herder (1744–1803) and Fichte (1762–1814). For Herder, a nation was synonymous with a language community, with each language community representing a unique variation of thought that separated it from other language communities, thus preserving a certain way of thinking over time. In the primordialist view, ethnic groups were characterized by a tight integration of geographic origin, language, territory, and religion. This also means that a people's language was intrinsically tied to its "race." Classifying human communities according to this view of ethnicity was widespread well into the twentieth century. The respective systems of classification – popularized, for example, in schoolbooks – tended to identify all Indo-Europeans or Semitic peoples with the White (or Caucasian) race, Tibeto-Burmese with the Mongoloid race, or speakers of Bantu languages with the "Negroid race" (Brues 1977). In the case of Finno-Ugrians and Northeast Africans, however, whole ethnolinguistic groups were simply declared either "Mongols" or "Caucasians," regardless of their appearance, implying supremacy of cultural traits over phenotypes.

## 2.5 Development: The Birth of Evolutionary Theory in the Enlightenment

Since the time of Darwin, ideas about development in general and evolution in particular have been reflected back and forth between the social and the biological domains. When Charles Darwin published his theory of descent with modification in 1859, he could draw on a tradition of evolutionary thinking in natural history and sociology that was maybe half a century old. Nevertheless, a basic model of evolution, meaning slow, gradual change from simpler to more complex forms, is even older, with its roots reaching back at least to the times of Enlightenment. According to Bock (1955, p. 133), it represents "a mode of conceiving change that is deeply rooted in Western thought." Bock (1955, pp. 129–30) points out that:

> The classical view of change as growth, the seventeenth century idea of progress, eighteenth century conjectural or hypothetical histories, and nineteenth century evolutionism all share in the perspective that change is natural, inevitable, slow, gradual and continuous. All depict change in terms of successive, finely graded stages of development. All seek in the original, from which change begins, the potential of what the changing thing is to become and the form which the process of change will take. All exclude consideration of specific events as effective factors in the process. All identify stages of evolution by reference to existing social or cultural forms that are arranged in a series according to a preconception of what evolution has actually been.

Good examples for Bock's exposition can be found in a number of prominent Enlightenment figures, specifically those from Scotland and France. It has been argued that in the eighteenth century, there was a special interests in the "comparative study of societies and how change in general takes place" in Scotland (Trigger 1998, p. 32), caused by the great divergence between the English-speaking, Protestant, economically developing cities and the rural, Gaelic-speaking, partly Catholic Highlands. Ferguson, in his Essay on the History of Civil Society (1767), conceives a developing natural world that is a home to progressively developing human societies. He describes progressive social development through three major stages, from savagery through barbarism to civilization. A stage model like this is quite typical for Enlightenment social thinkers, as is the progressive nature of development, brought mainly through technological advances, which in turn lead to social advances. Ferguson had been a professor of natural history at Edinburgh and, only 3 years before the publication of the essay, moved to a chair of moral philosophy. This might be one of the reasons for the close connections between natural and social development he constructs, but it also shows that disciplinary boundaries were less rigid in the eighteenth century than they are today.

Another example for proclaimed progressive stages of development is Condorcet's *Esquisse d'un tableau historique des progrès de l'esprit humain* (1795), in which he describes a progressive development of human societies through ten stages. All new stages are reached through progress in science and technology, and from this, progress in the moral and political sciences and then in social practice must follow. This would eventually lead to societies in which people increased both their mastery of nature and personal freedom. While short-term setbacks were possible, development at large remained progressive. Condorcet had begun the book before the French Revolution, and the strongly positivistic view of science, technology, and society that are apparent throughout his work is quite typical for the Enlightenment.

## 2.6 The Development of Sociology in the Nineteenth Century

The nineteenth century saw a proliferation of evolutionary theories. Not only did Lamarck (1809), Chambers (anonymously 1844), and Darwin (1859) publish the most important contributions to biological evolution up to these points in time, but evolutionism was firmly established as a school of thought in the human sciences, including sociology and anthropology.

Extending Concordet's view on the progressive development of mankind through stages, Comte (1798–1857), in his application of statistical methods and theoretical interpretations on societies, favored a strongly historical-orientated, evolutionary view on human societies. In the tradition of Quételet (1796–1874) who had examined the age of individual members of a society at certain intervals (Quételet 1838), Comte transferred the chronological aspect that is imbedded in the notion of "age" from individuals to a society as a whole. Comte introduced an historical perspective to the study of human societies by perceiving them as entities that had undergone "aging," i.e., development and change. For him, growth and gradual change in human communities constituted the main topic of studies in sociology. Even though Comte maintained the prevailing view that a society's characteristics were merely functions of its individuals' physical and biological features, he insisted on the fact that each society's unique structural features were caused by its specific history. To Comte, examining the history of a given society and drawing empirically based conclusions concerning social change constituted the main aim of research in sociology. These conclusions had then to be derived from the biological and physiological laws governing human nature (Fuchs-Heinrich 1998).

Seventeen years after Comte's voluminous Cours de philosophique positive (1830–1842) had been finished, Darwin (1809–1882) published his carefully prepared On the Origin of Species (1859), outlining his biology-based theory of evolution. That means that roughly two decades after Comte had presented a theory historicizing the development of human societies, Darwin historicized nature. This kind of evolution describes change as gradual and continuous and embraced the idea of the hereditability of acquired characteristics. We shall refer to it as Darwinian, and while Darwin held opinions on the process and material substance of inheritance, his theory does not, in its core, depend on them. The other important evolutionist in sociology was Herbert Spencer, who developed an evolutionary approach to sociology, which puts him in a tradition with Ferguson and Condorcet. Spencer employed Comte's term sociology to describe his study of human societies. He shared Comte's positivist view of science, but took it even further. While Comte had assumed an encompassing universality of scientific methods, Spencer assumed that a unity of all scientific knowledge would be possible and that all phenomena could be explained by one law: universality of evolution. In his *Progress: Its Law and Cause* (1857), Spencer develops a universal theory of evolution, which encompasses the cosmos at large, geology, living nature, humankind, and human societies. Spencer describes evolution as a development from simpler, homogenous toward more complex, more heterogenous forms. The close connection between social and biological theories of evolution in the mid-nineteenth century is illustrated by the close intellectual exchange between Spencer and Darwin: In the 1860s, Spencer adopted Darwin's "natural selection" into his *Principles of Biology* (1864–1867), while Darwin adopted Spencer's term "survival of the fittest" into the Origin of Species from the 5th edition (1869) onward.

## 2.7 *Development as Prelude to Civilization: Evolutionary Theories of Culture*

In Darwin's work, occasional references to cultural evolution can be found. In *Descent of Man*, he points to technology, such as traps, snares, or weapons, which might give one primitive human group a fitness advantage over another. In *The Origin of Species*, Darwin several times points to similarities between established theories of language change and his theory of descent with modification. In *Descent of Man*, he turns the argument around to show the wide applicability of his theory. He writes:

> A struggle for life is going on amongst the words and grammatical forms in each language. The better, the shorter, the easier forms are constantly gaining the upper hand. (Darwin 1877, p. 113)

Already before Darwin, Gustav Klemm (1802–1867), head of the Royal Library in Dresden, had presented a three-stage model of cultural evolution based on human social organization, technology, and belief systems in his *Allgemeine Kulturgeschichte der Menschheit* (1843–1852). Probably the most influential work of the nineteenth-century evolutionism in anthropology is Lewis Henry Morgan's *Ancient Society* (1877). Morgan was heavily influenced by the Bachofen (1815–1887) who – in *Das*

*Mutterecht* (1861) – had presented his view of an originally matriarchal society that anteced later evolutionary stages of patriarchal societies (Rössler 2007, p. 5). Like Ferguson and Klemm before him, Morgan distinguishes three broad stages ("savagery," "barbarism," and "civilization"), through which all human societies progress. Morgan divides the three stages into three substages each and uses both technological and social indicators to place societies in his system. On the lowest substage of savagery, people do not possess any subsistence technology and have to gather their food. On the highest substage of savagery, they develop bow and arrows and marriage between brother and sister becomes outlawed. Pottery and agriculture mark the lowest stage of barbarism, which is also associated with a more general incest taboo and the formation of clans and villages as the basic social unit. The highest substage of "barbarism" is characterized by the development of metal working the polygamous families. When writing and the monogamous family spread through a society, it has reached the first step and the lowest rank of civilization and will eventually progress toward a present-day Euro-American society.

## 3 Degeneration

### *3.1 Fear of Degeneration by Admixture*

As a conclusion, we might point out that – in the nineteenth-century evolutionist's view – evolution works according to a pedigree model, involving vertical transfer. Lateral or horizontal transfer, on the other hand, does not contribute to development to higher stages on the evolutionary chain, but to an evolutionary backlash, i.e., degeneration, spoiling the lineage of the evolutionary tree. The discourse on degeneration by admixture of both "races" and languages presumes an acceptance of the possibility of lateral transfer. This is illustrated by the fact that fear of degeneration was a topic of concern in both racial theory and linguistics. Some evolutionary theories, especially those of social evolution, were progressive and assumed that later forms would be more perfect than earlier ones. Spencer's universal evolutionary theory is an example for this view of the evolutionary process. Other theories assumed that earlier forms were closer to a pure, unspoiled primal form, while later ones could be contaminated by contact with lesser forms. At this point, a short overview concerning the history of the term degeneration may be useful.

### *3.2 The History of the Term "Dégéneresence": Pure and "Primitive" Races*

The term "dégénérescence" existed before his most prominent promoter Bénédict-Augustin Morel (1809–1875) popularized the term (Morel 1857). Jacques-Joseph Moreau de Tours (1804–1884), for example, used this term in 1855 to describe the phenomenon that interbreeding in animals may lead to defective offspring, a clear indication that lateral transfer in biology was regarded as potentially hazardous to the sane development of species (Moreau de Tours 1855).

After Moreau de Tours, however, the term degeneration acquired a somewhat different meaning. Morel not only delineated something primarily medical but integrated his religious beliefs into his concept of degeneration. He outlined this concept in his *Traité des dégénérescence physiques, intellectuelles et morales de l'espèce humaine et de ses causes qui produisent ces varieties maladives* (1857) which he wrote explicitly as a preparation for his textbook on mental diseases. This treatise aims to describe the origin of pathological changes in human beings in relation to biblical Genesis. In the beginning there is a pristine-"type primitive" that represents Adam before his fall. Thus, mankind according to Morel is not a product of previous transformations but existed sui generis. Degeneration and illness are deviations from this "type primitive," originating from external, environmental, and social circumstances. After his fall, man is unable to escape from these circumstances, which results in hereditary deviations. Though the causes of degeneration are external, its origin lies in man's original sin.

Deviations lead in two different directions. One is the development of (normal) "human races"; the other is the development of morbid physical, mental, or moral traits within these races, the latter process being called degeneration by Morel. Although both developments differ from the "type primitive," "normal races" still follow divine law. Physical and moral degeneration is heritable and progressive until extinction. Degeneration is correlated to pathological anatomy, and therefore, degeneration can be classified according to morphologic-anthropometric criteria following external stigmata.

## *3.3 Degeneration in Linguistics: "Mixed and Spoiled Languages"*

The origin of the term degeneration makes clear that the concept was primarily developed and used in the medical field, but by referring to Schleicher (1850), we have already seen that the term degeneration provides a fine example for the transfer of concepts between sciences (in this case medicine as a basis of physical anthropology) and the humanities and occurred in linguistics as well.

In the prevailing tradition of historical comparative linguistics, it was frequently argued that the original, pure Indo-European language was represented by Sanskrit (Jones 1786) and that Indo-Iranian (Aryan) languages were the purest forms alive today. All the other branches of the Indo-European family had been corrupted due to migration and intense contact with other non-European languages (Schlegel 1808; Schleicher 1850; Müller 1855). This view illustrates that the search for the pristine, original form of language has been a constant topic in nineteenth-century linguistics and further shows that vertical transfer between languages was considered to be a reason for deleterious language change and modification.

The evolutionary scheme of language typology also took into account the possibility for languages to revert to a lower level and thus degenerate. Schleicher employed the notion of degeneration to describe non-evolutionary development of languages by convergence and diffusion. He considered similarities between Balkan languages that constituted a union of only distantly related languages the result of degeneration and corruption, contrary to the supposedly pure genealogy of the rest of the Indo-European languages:

> It is a remarkable fact that at the lower Danube and further southwest, a bunch of neighboring languages is to be found which – apart from their different origin – only share one feature in the sense that they are the most spoiled representatives of their respective families. These wayward sons of languages are Vlah in the Romance family, Bulgarian in the Slavonic family, and Albanian in the Greek family of languages. The degeneration has achieved a minor degree in the northernmost language (mentioned at first); a higher degree in the central one, Bulgarian; and a degree almost completely obscuring its origin in the southernmost language, i.e., Albanian.[4]

## 3.4 *The Myth of the "Pure" Aryan Race and Language*

In biology, the idea of degeneration was mainly applied to "races." At about the same time that modern nation states started to be formed, national populations became equated with distinct races, and strict endogamy was proposed as a policy to keep populations pure. The idea that a people's cohesion, health, and well-being were intrinsically linked to its presumed "racial qualities" and that racial mixture would constitute an offense against racial purity had been developed in the course of the nineteenth century. Originating with Schlegel (1808) who claimed that linguistic affiliation provided better evidence concerning the purity of a population's heritage (pureté du sang) than comparing physical features, language started to be seen as an indicator of race – a view that was perpetuated by Pictet (1859–1863), Gobineau (1853/1854), Le Bon (1894), Vacher de Lapouge (1899), and Günther (1934), among others.

The emphasis on "racial purity" might well be a reflection of early colonial experiences since the expansion of the European powers had raised the awareness of "racial mixture" of colonizers and colonized which was discouraged out of fear of degenerating "higher races." Of course, this policy remained a mere theory, and

[4] Original quote: "Es ist eine bemerkenswerte Erscheinung, daß um die Untere Donau und weiter nach Südwesten sich eine Gruppe aneinandergrenzender Sprachen zusammengefunden hat, die beim stammhafter Verschiedenheit nur darin übereinstimmen, daß sie die verdorbensten ihrer Familien sind. Diese mißratenen Söhne sind das Walachische in der romanischen, das Bulgarische in der slavischen und das Albanische in der griechischen Familie. Das Verderbnis zeigt in der nördlichsten Sprache, der zuerst genannten, noch in einem geringen Grade, mehr schon in der mittleren, dem Bulgarischen, und hat in der südlichen, der albanesischen einen ihrer Herkunft fast völlig verdunkelnden Grad erreicht." (Schleicher 1850, p. 143)

"mixed offspring" became a common sight in all colonies. Thus, the proponents of endogamy had always been aware of human consanguinity.

When, in the course of the nineteenth century, "races" tended to be increasingly associated with certain language families, the notion that the speakers of Indo-European languages originated from an ancestral race that was defined by geographic, cultural, biometric, and linguistic similarity was propagated by authors like Schlegel (1808), Pictet (1859–1863), Gobineau (1853–1855), Le Bon (1894), and Vacher de Lapouge (1899). It was Lapouge who claimed that the "Nordic race" was the ideal model of the superior "Aryans" who could be identified biologically by measuring their cephalic index, representing the long-headed (dolichocephalic) blond Europeans who were natural leaders to rule over more brachiocephalic (short-headed) people. With Le Bon (1894) who simply identified the "higher races" with Indo-Europeans, we witness the transfer of a linguistic term into physical anthropology. Henceforth, scholars who took a primordialist perspective assumed a direct, intrinsic connection between physical appearance, culture, and language.

Combining archaeological, linguistic, and biometric data, the idea developed in the twentieth century that speakers of the proto-Indo-European language were related to the late Neolithic *Corded Ware Culture* of Central Europe (Kossina 1902; Hirt 1905; Childe 1926; Günther 1934, pp. 16–17) and to the assumed "Nordic race." The superiority of this presumed race was attributed to natural selection – in a Darwinian sense – which had supposedly taken place due to harsh environmental conditions in the late Pleistocene and postglacial periods (Günther 1934). Thus, it was claimed that the "Nordic race" represented the earliest and "purest" speakers of Indo-European. It was claimed that, while some of the stock had degenerated due to migration and racial admixture, for example, in the case of the Aryans who settled in India from the second millennium B.C. onward, the pure stock had been best preserved by Germanic-speaking peoples. Some of these ideas provided a strong ideological backing for racism and National Socialism in the first half of the twentieth century (Römer 1989).

## 3.5 "Degeneration in Science": The Decline of Evolutionism

When comparing nineteenth and twentieth centuries' attempts to classify human linguistic and phenotypic diversity, it becomes clear that certain traditions prevail. In the late eighteenth and nineteenth centuries, populations had been defined by an integrated complex of cultural traits (including language, religion, and nationality) according to a primordialist view. In the beginning of the twentieth century, this view met fierce opposition by the emerging cultural relativism and particularism of the Boasian school of cultural anthropology. In opposition to the prevailing ethnocentric orientation of his time, Boas strongly rejected speculative ideas of cultural evolution and claimed that culture developed independently of biological characteristics of human populations, with culture, "race," and language constituting mutually independent and unrelated determinants of human existence. Boas and his North American disciples did not only deny the idea of coevolution of human languages, cultures, and physical traits. They dispensed with the notion of cultural evolution altogether, since culture did not develop according to a hierarchy of successive predetermined stages as outlined by Morgan (1877), Tylor (1871), or one of their eighteenth-century predecessors. Instead, each culture constituted its own equilibrium of thought and feeling and the labeling cultures "primitive" only made sense in regard to their specific technical equipment (Streck 2000, p. 142). Looking beyond the development of U.S. American cultural anthropology, it can be concluded that ethnological and sociological evolutionism was dismissed with the establishment of social anthropology in Britain (Malinowski 1915; Radcliff-Brown 1922; Evans-Pritchard 1937) and France (Durkheim 1912; Mauss 1913) where it later was coined structuralism (*anthropologie structurale*, cf. Lévi-Strauss 1958, 1967).

In the German-speaking countries, cultural diffusionism took an equally critical stand toward the question of cultural evolution in the shape of "Kulturkreislehre" (Frobenius 1898; Ankermann 1905; Graebner 1905, 1911) elaborated on by the "Vienna School" of historical ethnography (Schmidt 1912–55, Koppers 1915–16). Anthropological historicism is the idea that different cultures emerge from a common predecessor, but each adapts to their unique geographic and historical environments. It emerged at a time when the idea of overall progress was increasingly questioned (Streck 2000, p. 142) and had been advocated by the geographers Georg Gerland (1833–1919) and Friedrich Ratzel (1844–1938, Ratzel 1885) in their application of Moritz Wagner's (1813–1887) idea of diffusion. Wagner's field of interest covered natural history, zoology, and geography as well as ethnography. As a matter of fact, the zoologist and geographer Ratzel was mentioned by Boas as one of his most influential early mentors in 1911 (Voget 1970, p. 209). In accordance with the traditions of German historicism, diffusionist ethnography was searching for "pure forms of culture" instead of a development in hierarchical steps (Streck 2000, p. 143), assuming that major technical and cultural inventions occurred only very rarely and were transmitted by cultural diffusion rather than by evolution – a view that prevailed in the work of Boas and his early disciples (Wissler 1926; Kroeber 1939, 1940) as well. Furthermore, instead of cultural evolution in progressive stages, the lore of "Kulturkreislehre" advocated the idea of cultural degeneration which was supposedly manifested in the supposed decline of monotheism to polytheism (Schmidt 1912–55) or the historical development of *Primärkultur* and *Sekundärkultur* as degenerative process spoiling features like monogamy, monotheism, and patriarchal structures that were still abundant in the assumed *Urkultur* (Rössler 2007, p. 13). This idea of decay, so prominent in fin de siècle thought in the German-speaking countries of central Europe (Spengler 1918), proved to have a lasting influence on German and Austrian ethnography (*Völkerkunde*) until the 1930s and 1940s (Rössler 2007).

### *3.6 The Reemergence of Evolutionism in the "Sciences of Man": Back to Unilinear Descent and Consecutive Development in Stages?*

Apart from influences of historicist thinking that can be clearly traced in Boasian anthropology, Boas is also known as the father of the four field approach in anthropology. He was a key figure in integrating sociocultural, linguistic, biological, and archaeological perspectives into a disciplinary framework that to this day fosters the dialogue between scholars who are inclined toward natural sciences with those inclined toward the humanities. In this way, Boas and those in his tradition, who initially opposed cultural evolutionism, eventually contributed to the reemergence of ideas of coevolution in biodiversity and linguistic diversity that occurred in the last quarter of the twentieth century in the context of U.S. American Anthropology. Theories of coevolution were predated by neo-evolutionists anthropologists like White (1900–1975), Steward (1902–1972), and Sahlins (born in 1930). With them, the idea of evolution had received renewed interest in cultural anthropology since the early 1950s (White 1959; Steward 1955; Sahlins and Service 1960). But these representatives of the neo-evolutionism restricted their evolutionary approach to cultural and social phenomena, not linking models of a unilinear (White) or multilinear (Steward) development of cultures and societies in successive stages to the evolution of human linguistic and biological diversity. Cultural evolutionism in the second half of the twentieth century is thoroughly informed by the modern-day evolutionary paradigm in biology, which includes a distinct medium of heredity and the medium's properties and mechanic's influence heredity itself. Also, change is not always described as gradual, but will occasionally occur in burst, especially in small populations. This model of cultural evolution shall be referred to as neo-Darwinian. Scholars of this period are concerned with the relationship between biological evolution and cultural evolution as two distinct processes following similar rules to explain human behavior and societies.

The emergence of the idea of unilinear human evolution that we outlined before was closely connected to a tradition of constructing tree models of both the biological and the linguistic evolution of humans which, too, can be traced back to the eighteenth century. In both physical anthropology and linguistics, treelike models of descent concerning languages and human biodiversity proved to be equally attractive until the present day. This resulted in repeated attempts to combine phylogenetic trees of human languages and population to construct a universal model of human descent – a research agenda that can be traced through the nineteenth and twentieth centuries. Thus, a new synthetic approach intending to present a global phylogeny of mankind was presented from the 1980s onward, being inspired by Dawkins (1976) coining the word *meme* to describe cultural replicators and representing a Renaissance of primordialist ideas in the shape of the "new synthesis" between genetic, linguistic, and archaeological data during the 1980s (Cavalli-Sforza et al. 1988; Renfrew 1987; Ruhlen 1987). Attempting to identify a connection between linguistic macro-phyla and genetic clusters of humankind to form a universal human pedigree or global phylogeny, this research program constitutes a more recent example to trace the origin of human languages and human biodiversity by an evolutionary approach. Interestingly enough, all fields of the Boasian four field approach are represented in this research agenda:

Physical anthropology, in the form of population genetics, is represented by Barbujani (1991, 1997), Cavalli-Sforza et al. (1988), and Cavalli-Sforza and Seielstad (2001) and elaborated by Greenberg (1963, 1971, 1987) and Renfrew (1987).

Linguistics is represented by long-range comparison. Joseph Greenberg was the most prominent linguist who employed mass comparison and the concept of basic vocabulary outlined by the propagators of lexicostatistics in the so-called "super grouping" approach to find macro-phyla of languages (Greenberg 1963, 1971, 1987, 2000, 2002).

Cultural anthropology is represented by more recent examples for an attempted synthesis of cultural and linguistic evolution by Gray (2005), Gray and Atkinson (2003), Gray et al. (2009), and Mace and Holden (2005). In their recent (2009) article, Gray et al. identify "the pleasures and perils of Darwinizing culture (with phylogenies)" and outline the promises and problems of this approach in detail. They state that language seems to be an especially suitable field to apply evolutionary explanations to. Finding such a connection would lead credibility to primordialist thinking in the sense that a close link between the vertical transfer of human genes and languages would be implied. Horizontal transfer and its impact on phylogenetic trees is a major concern for the authors who point out that "many of the traits of most interest to anthropologists involve codified practices and ancient rituals with tighter integrational constraints that are likely to limit the impact of horizontal transfer" (Gray et al. 2009, pp. 15–16).

Archaeology is presented by Colin Renfrew in his archaeogenetics approach, attempting to trace prehistoric human migrations by the use of genetic analysis und long-distance linguistic comparison and supporting the "Out of Anatolia" hypothesis concerning the origin of the Indo-European language family and its speakers (Renfrew 1987).

All these scholars form a scientific community (in the sense of Kuhn 1962) within anthropology, sharing an evolutionary approach to their work. The research program of "global phylogeny" does not necessarily imply the idea of cultural, linguistic, or physical degeneration by ad mixture (vertical transfer) of genes, language features (phonemes, morphemes), or memes, but emphasizes vertical transfer of languages, culture, and genes as favored and prevailing feature of human evolution. Furthermore, this model strongly supports the idea of human linguistic, cultural, and biological coevolution. The research program of the so-called new synthesis has been criticized from both cultural anthropologists and comparative linguists (e.g., Kressing 1994, 2012; Kressing et al. 2013; Marks 1994, 1995; McMahon 1995). Cultural anthropologists point to the fact that factors such as linguistic macro-families and genetic clusters only sometimes, but by no means always, match up, that great cultural diversity is found within macro-families, and that a direct correlation between genetic clusters and linguistic macro-phyla is not always testified.

Comparative linguists' criticism is even more fundamental: They attack the data sets, mainly based on the Swadesh Lists used for lexicostatistical and glottochronological comparison from the 1950s onward that many of the linguistic macro-families are based on claiming chance resemblances and cultural bias of these lists. The model of "global phylogeny" does only in a marginal way take into account linguistic assimilation processes, e.g., the role of substrata and superstrata, thereby focusing on a primordialist conception of ethnicity and perpetuating the ethnographical myth of indigenous peoples having lived in isolated communities of primitive hunters and gatherers or agriculturalists for hundreds or thousands of years without interethnic relations contributing to gene flow and linguistic borrowings (Marks 1994, pp. 176–177).

## 4 Conclusion

Apart from the research agenda in the search for "global phylogenies" outlined above, cultural anthropology and linguistics on one side (as part of the humanities) and physical anthropology as a biological science on the other seem to belong to two different cultures today: the sciences and the humanities. Yet, evolutionary approaches, which have been very successful in the sciences, were increasingly applied to the study of culture, starting with neo-evolutionism in cultural anthropology in the 1950s. In a recent publication, Sahlins (2000) also shows how the evolutionary theoretical framework was reintroduced into history and sociology in the 1970s. The theories of evolution that seem most applicable to the study of culture are based on an abstract theory of evolution, like Lewontin's (1970) three steps of phenotypic variation, differential fitness, and heritability of fitness variability. In cultural evolution, vertical transmission (like language acquisition of children from their parents) plays an important role, but so do horizontal and oblique transmission. One might even imagine a scenario in which vertical transmission is completely replaced by oblique transmission (and even the terms themselves become questionable), like an orphan who grows up in a foster family and acquires language from his foster parents. Another factor that makes increasingly neo-Darwinian models of cultural evolution problematic is the heritability of acquired characteristics, which is not part of the neo-Darwinian paradigm, while it is of great importance in cultural transmission and we would expect that once a person has learned a skill, she/he will be able to teach it to others. For these reasons, cultural phylogenies do not prove to be very convincing, and a reticulate approach can be regarded to be much more appropriate to illustrate processes of cultural transmission and development.

Analyzing the historical development in anthropology, we have demonstrated that in the eighteenth and nineteenth centuries, no sharp distinction was made between the two scientific communities of racial theorists and linguistics who both imagined the development of mankind through similar stages, assuming that intellectual, technical, and moral progress were bound to one another in these stage models. In primordialist models of developments, language, culture, and race were intrinsically linked. We have also shown that the idea of degeneration was present in theories of development of species, populations, and languages. Concomitantly with the emergence of evolutionary thought envisaging progressive stages of development, the idea of degeneration and decay played a major role in anthropological thought until the 1950. In our opinion, it was the shared fear of pristine forms being spoiled by admixture that contributed to the discourse on degeneration and was used to justify racial separation and attempts to maintain national, racial, and linguistic purity.

We have shown that the history of anthropology is deeply intertwined with that of its contributing disciplines, which now form the branches of physical and cultural anthropology. Historically, a network of scholars and ideas between representatives of these scholarly traditions of "humanid" and "hominid" anthropology (Streck 2000) has been maintained, even though periods of increasing institutional separation. As a whole, the history of classification and evolution of human populations and languages shows a constant back and forth between scholarship concerned with the domains of culture and biological nature. While the transfer of models and methods is potentially problematic, the exchange has often been, and still has the potential to be, fruitful for both sides.

## References


Ankermann, B. (1905). Kulturkreise und Kulturschichten in Afrika. *Zeitschrift für Ethnologie, 37*, 54–90.

Arroux, S. (1990). Representation and the place of linguistic change before comparative grammar. In T. De Mauro, & L. Formigari (Eds.), *Leibniz, Humboldt, and the origins of comparativism* (pp. 231–238). Amsterdam: J. Benjamins.

Bachofen, J. J. (1861). *Das Mutterecht*. Basel: Beno Schwabe.

Barbujani, G. (1991). What do languages tell us about human microevolution? *Trends in Ecology and Evolution, 6*(5), 151–156.

Barbujani, G. (1997). DNA variation and language affinities. *American Journal of Human Genetics, 61*, 1011–1014.

Bastian, A. (1860). *Der Mensch in der Geschichte*, (3 vol.). Leipzig: Wigand.

Beddoes, T. (1799). *Contributions to physical and medical knowledge, mainly from the West of England*. Bristol: Biggs & Cottle.

Blumenbach, J. F. (1775). *De generis humani varietate nativa liber*. Göttingen: Rosenbusch.

Boas, F. (1913). *Kultur und Rasse*. Berlin: Gruyter.

Boas, F. (1940). *Race, language, culture*. New York: MacMillan.

Bock, K. (1955). Darwin and social theory. *Philosophy of Science*, 22(2), 123–134.

Bopp, F. (1816). *Über das Conjugationssystem der Sanskritsprache in Vergleichung mit jenem der griechischen, lateinischen, persischen und germanischen Sprache*. Hildesheim: Olms [reprint].

Bowler, P. (1983). *Evolution: The history of an idea*. Berkeley: University of California Press.

Brues, A. M. (1977). *People and races*. New York: MacMillan.

Buffon, G.-L. (1749–1804). *Histoire naturelle générale et particulière* (44 Vol.). Paris: Imprimeries royale, Plassan.

Burdach, F. (1800). *Propädeutik zum Studium der gesammten Heilkunst*. Leipzig: Breitkopf & Härtel.

Carus, C. G. (1849). *Über die ungleiche Befähigung der verschiedenen Menschenstämme für höhere geistige Entwicklung*. Leipzig: Brockhaus.

Cassmann, O. (1594). *Psychologia anthropologica, sive animae humanae doctrina*. Hanau.

Cavalli-Sforza, L. L., & Seielstad, M. (2001). *Genes, peoples, and languages*. Berkeley: University of California Press.

Cavalli-Sforza, L. L., Piazza, A., Menozzi, P., & Mountain, J. (1988). Reconstruction of human evolution: bringing together genetic, archaeological, and linguistic data. *Proceedings of the National Academy of Sciences of the United States of America, 85*, 6002–6006.
Chambers, R. [anonymously] (1844). *Vestiges of the natural history of creation*. London/Edinburg: W. & R. Chambers.
Childe, V. G. (1926). *The Aryans: A study of Indo-European origins*. London: K. Paul.
Comte, A. (1830–42). *Cours de philosophie positive*. Paris: Librairie Larousse.
Condorcet, M. J. A. de. (1795). *Esquisse d'un tableau historique des progrès de l'esprit humain*. Paris: Agasse.
Darwin, C. (1859). *On the origin of species by means of natural selection, or the preservation of favoured races in the struggle for life*. London: John Murray [5th ed. 1869].
Darwin, C. (1871) [1881]. *The descent of man, and selection in relation to sex*. London: John Murray.
Dawkins, R. (1976). *The selfish gene*. Oxford: Oxford University Press.
de Saussure, F.. (1916). *Cours de linguistique générale*. Paris, Lausanne: Payot.
Descartes, R. (1641). *Meditationes de prima philosophia*. Paris: M. Soly.
Desmond, A., & Moore, J. A. (2009). *Darwin's sacred cause. Race, slavery and the quest for human origins*. London: Allen Lane.
Deutsche Gesellschaft für Völkerkunde, Berliner Gesellschaft für Anthropologie, Ethnologie und Urgeschichte (Eds.), *Zeitschrift für Ethnologie*. Berlin: Reimer (first publ. 1869).
Driem, G. van (2005). Sino-Austronesian vs. Sino-Caucasian, Sino-Bodic vs. Sino-Tibetan, and Tibeto-Burman as default theory. In Y. P. Prasada, & Bh. Y. Govinda, et al. (Eds.). *Contemporary issues in Nepalese linguistics* (pp. 305–338). Kathmandu: Linguistic Society of Nepal.
Driem, G. van (2007). The diversity of the Tibeto-Burman language family and the linguistic ancestry of Chinese. *Bulletin of Chinese Linguistics*, 1(2), 211–270.
Durkheim, E. (1912). *Les formes élémentaires de la vie religieuse*. Paris: F. Alcan.
Evans-Pritchard, E. (1937). *Witchcraft, oracles and magic among the Azande*. Oxford: Oxford University Press.
Fangerau, H., Krischel, M., & Kressing, F. (2009). Computergestützte Analyse in Biologie, Sprach- und Geschichtwissenschaft. In St. Fischer, E. Maehle, & R. Reischuk (Eds.), *Informatik 2009: Im Focus das Leben, Beiträge der 39. Jahrestagung der Gesellschaft für Informatik e.V.* (GI), 28.9.-2.10.2009, Lübeck, Proceedings (pp. 582–594). Lübeck.
Fangerau, H., Geisler, H., Halling, T., & Martin, W. F. (2013) (Eds.). *Classification and evolution in biology, linguistics and the history of science. Concepts – methods – visualization*. Stuttgart: Steiner.
Ferguson, A. (1767). *Essay on the history of civil society*. London/Edinburgh: Tadell, Creech & Bell.
Forster, G. (1786). Noch etwas über die Menschenrassen. *Teutscher Merkur, 56*, 57–86.
Frazer, J. G. (1890). *The golden bough: A study in magic and religion*. London: Macmillan. [2nd edition].
Frobenius, L. (1898). *Ursprung der afrikanischen Kulturen*. Berlin: Bornträger.
Fuchs-Heinrich, W. (1998). *Auguste Comte. Einführung in Leben und Werk*. Hagener Studientexte zur Soziologie 2. Opladen: Westdeutscher Verlag.
Gobineau, A. (1853/1855). *L'essai sur l'inégalité des races humaines*. Paris: Firmin-Didot.
Graebner, F. (1905). Kulturkreise und Kulturschichten in Ozeanien. *Zeitschrift für Ethnologie, 37*, 28–53.
Graebner, F. (1911). *Methoden der Ethnologie*. Heidelberg: Winter.
Gray, R. D. (2005). Pushing the time barrier in the quest for language roots. *Science, 309*, 2007–2008.
Gray, R. D., & Atkinson, Q. D. (2003). Language-tree divergence times support the Anatolian theory of Indo-European origin. *Nature, 426*, 435–439.
Gray, R. D., Greenhill, S. J., & Ross, R. M. (2009). The pleasures and perils of Darwinizing culture (with phylogenies). *Biological Theory, 2*(4), 360–375.
Greenberg, J. H. (1963). *The Languages of Africa*. The Hague/Bloomington: Mouton de Gruyter, Indiana University Center.
Greenberg, J. H. (1971). The Indo-Pacific Hypothesis. In T. Seboek, et al. (Eds.) *Current Trends in Linguistics, 8* (pp. 807–871). The Hague: Mouton de Gruyter.

Greenberg, J. H. (1987). *Language in the Americas.* Stanford: Stanford University Press.
Greenberg, J. H. (2000). *Indo-European and Its Closest Relatives: The Eurasiatic Language Family, 1, Grammar.* Stanford: Stanford University Press.
Greenberg, J. H. (2002). *Indo-European and Its Closest Relatives: The Eurasiatic Language Family, 2, Lexicon.* Stanford: Stanford University Press.
Grimm, J. (1819–1834). *Deutsche Grammatik.* Göttingen: Dieterich.
Günther, H. F. K. (1934). *Die Nordische Rasse bei den Indogermanen Asiens. Zugleich ein Beitrag zur Frage nach der Urheimat und der Rassenherkunft der Indogermanen.* München: Lehmann.
Hann, C. (2005) (Ed.). *One discipline, four ways: British, German, French, and American anthropology. The Halle Lectures.* Chicago, London: University of Chicago Press.
Herder, J. G. (1772). *Abhandlung über den Ursprung der Sprache.* Berlin: Voß.
Hirt, H. (1905–1907). *Die Indogermanen. Ihre Verbreitung, ihre Urheimat und ihre Kultur* (2 Vol.). Strassbourg: Trübner.
Hoßfeld, U. (2005). *Geschichte der biologischen Anthropologie in Deutschland.* Stuttgart: Steiner.
Humboldt, W. (1836). *Über die Verschiedenheit des menschlichen Sprachbaus und ihren Einfluß auf die geistige Entwicklung des Menschengeschlechts.* Berlin: Kgl. Akad. D. Wiss., F. Dümmle.
Hundt, M. (1501). *Anthropologium de hominis dignitate, natura et proprietatibus.* Leipzig: Wolfangum.
Jakobson, R. (1931). Über die phonologischen Sprachbünde. In R. Jakobson (Ed.), *Selected writings. Vol. 1: phonological studies* (pp. 137–143). 's-Gravenhage: Nijhoff.
Jones, W (1786). The third anniversary discourse, on the Hindus, delivered by the president, February 2, 1786. *Asiatic Researches, 1,* 415–431.
Kant, I. (1775). *Von den verschiedenen Rassen der Menschen.* Berlin: Hartung.
Kant, I. (1785). Zur Bestimmung des Begriffs der Menschenrasse. *Berlinische Monatsschrift, 6,* 390–417.
Klemm, G. F. (1843). *Allgemeine Culturgeschichte der Menschheit.* Leipzig: Teubner.
Koppers, W. (1915–16). Die ethnologische Wirtschaftsforschung: Eine historisch-kritische Studie. *Anthropos, 10,* 611–651, 11, 971–1079.
Kossina, G. (1902). Die indogermanische Frage archäologisch beantwortet. *Zeitschrift für Ethnologie, 34,* 161–222.
Kressing, F. (1994). Das 'Human Genome Diversity Project' – Rassismus im neuen Gewand oder erkenntnistheoretische Grundlagenforschung zum Wohle der Menschheit? *INFOE-Magazin, 1,* 16–21.
Kressing, F. (2012). Screening Indigenous peoples' genes – the end of racism or postmodern bioimperialism? In S. Berthier-Forglar, S. Tolazzi, & S. Collingwood-Whittick (Eds.), *Biomapping indigenous peoples. Towards an understanding of the issues* (pp. 117–136). Amsterdam, New York: Rodopi.
Kressing, F. (2013). Mapping human biological and linguistic diversity – a bridge between sciences and humanities. In H. Fangerau, H. Geisler, T. Halling, & W. F. Martin (Eds.), *Classification and evolution in biology, linguistics and the history of science. Concepts – methods – visualization* (pp. 97–108). Stuttgart: Steiner.
Kressing, F., Krischel, M., & Fangerau, H. (2013). The 'global phylogeny' and its historical legacy: A critical review of a unified theory of human biological and linguistic co-evolution. *Medicine Studies.* doi:10.1007/s12376-013-0081-8.
Krischel, M., & Fangerau, H. (2013). Historical network analysis can be used to construct a social network of 19th-century evolutionists. In H. Fangerau, H. Geisler, T. Halling, & W. F. Martin (Eds.), *Classification and evolution in biology, linguistics and the history of science. Concepts – methods – visualization* (pp. 45–65). Stuttgart: Steiner.
Krischel, M., & Kressing F. (2014). Netzwerke statt Stammbäume? Lateraler Transfer in Evolutionstheorien von Sprachen, Arten und Kultur. In Deutsche Gesellschaft für Geschichte und Theorie der Biologie (Ed.), *Ordnung – Organisation – Organismus. Beiträge zur 20. Jahrestagung der DGGTB.* Verhandlungen zur Geschichte und Theorie der Biologie, 18, (pp. 103–116). Berlin: Verlag für Bildung und Wissenschaft.
Krischel, M., Kressing, F., & Fangerau H. (2011). Die Entwicklung der Deszendenztheorie in Biologie, Linguistik und Anthropologie als Austauschprozess zwischen Geistes- und

Naturwissenschaften. In H.-K. Keul, & M. Krischel (Eds.), *Deszendenztheorie und Darwinismus in den Wissenschaften vom Menschen* (pp. 107–121). Stuttgart: Steiner.
Kroeber, A. L. (1939). *Cultural and natural areas of Native North America*. Berkeley: University of California Press.
Kroeber, A. L. (1940). Stimulus diffusion. *American Anthropologist, 42*, 1–20.
Kuhn, T. H. (1962). *The structure of scientific revolutions*. Chicago: University of Chicago Press.
Lamarck, J. -B. (1802). *Hydrogéologie*. Paris: Agasse.
Lamarck, J. B. (1809). *Philosophie zoologique, ou, exposition des considérations relative à l'histoire naturelle des animaux*. Paris: Dentu.
Le Bon, G. (1894). *Lois psychologiques de l'évolution des peuples*. Paris: Alcan.
Lévi-Strauss, C. (1958). *Anthropologie structurale*. Paris: Libraire Plon.
Lévi-Strauss, C. (1967). Structural analysis in linguistics and anthropology. In A. Collins (Ed.), *Structural anthropology* (pp. 66–79). New York: H. Fertig.
Lewontin, R. C. (1970). The units of selection. *Annual Reviews of Ecology and Systematics, 1*, 1–18.
Linnaeus [Linné], C. (1735). *Systemae Naturae*. Leiden: De Groot.
Mace, R., & Holden, C. J. (2005). A phylogenetic approach to cultural evolution. *Trends in Ecology and Evolution, 20*(3), 116–121.
Malinowski, B. (1915). *The Trobriand Islands*. London: Routledge.
Malte-Brun, C., Lapie, P., & Poirson, J. -B. (1810–1826). *Précis de la géographie universelle, ou description de toutes les parties du monde sur un plan nouveau, d'après les grandes divisions naturelles du globe: Collection de cartes géographiques dirigées par M. Malte-Brun, dressées par MM. Lapie et Poirson* (8 Vols.). Paris: François Buisson.
Marks, J. (1994). What is the background of the human genome diversity project? http://personal.uncc.edu/jmarks/hgdp/hgdp1.html. Accessed 4 Oct 2010.
Marks, J. (1995). *Human biodiversity. Genes, race, and history*. New York: Aldine de Gruyter.
Mauss, M. (1913). L'ethnographie en France et a l'etranger. *Revue de Paris, 20*(5), 537–560, 815–837.
McMahon, A. (1995). *Understanding language change*. Cambridge: Cambridge University Press.
Moreau de Tours, J. -J. (1855). De l'identité de l'état de rêve et de la folie. *Annales médicopsychologiques*, 3(1), 361–408.
Morel, B. -A. (1857). *Traité des dégénérescences physiques, intellectuelles ou morales de l'espèce humaine et des causes qui produisent ces variétés maladives*. Paris: J.-B. Baillière.
Morgan, L. H. (1877). *Ancient Society, or: researches in the lines of human progress from savagery through barbarism to civilization*. New York: Holt.
Müller, F. M. (1855). *Languages of the seat of war in the East, with a survey of the three families of language, Semitic, Arian, and Turanian*. London: Williams & Norgate.
Oken, L. (1809–1811). *Lehrbuch der Naturphilosophie*. Jena: Frommann.
Pictet, A. (1859–1863). Les origins indo-européenes ou les Aryas primtifs. Essai de paléontolgoie linguistique (2 Vol.). Paris: Cherbuliez.
Postel, G. (1538). *Linguarum duodecim characteribus differentium alphabeta, introductio ac legendi modus longe facilimus*. Paris: Dionysium Lesenier.
Quételet, A. (1838). *Mortalité aux différents ages dans la Belgique*. Bruxelles: M. Hayez, Emprimeuer de l'academie royale.
Radcliff-Brown, A. R. (1922). *The Andaman Islanders. A study in Social Anthropology*. London: Cambridge University Press.
Rask, R. (1818). *Undersøgelse om det gamle nordiske eller islandske Sprogs Oprindelse*. Købnhavn: Gyldendål.
Ratzel, F. (1882–1891). *Anthropogeographie – Die geographische Verbreitung des Menschen*. Stuttgart: J. Engelhorn.
Ratzel, F. (1885). *Völkerkunde*. Leipzig: Bibliographisches Institut.
Renfrew, C. (1987). *Archaeology and language: The puzzle of Indo-European origins*. Cambridge: Cambridge University Press.
Römer, R. (1989). *Sprachwissenschaft und Rassenideologie in Deutschland*. München: Fink.
Rössler, M. (2007). *Die deutschsprachige Ethnologie bis ca. 1960: Ein historischer Abriss*. Cologne: Institute of Cultural and Social Anthropology, Working Papers No. 1.

Ruhlen, M. (1987). *A Guide to the World's Languages*. London, Melbourne, Auckland.
Sahlins, M. (2000). Ethnographic experience and sentimental pessimism: Why culture is not a disappearing object. In L. Daston (Ed.), *Biographies of scientific objects* (pp. 158–202). Chicago, London: University of Chicago Press.
Sahlins, M., & Service, E. R. (1960). *Evolution and culture*. Ann Arbor: University of Michigan Press.
Sapir, E. (1921). *Language. An introduction to the study of speech*. New York: Harcourt Brace.
Schlegel, F. (1808). *Ueber die Sprache und Weisheit der Indier. Ein Beitrag zur Begründung der Alterthumskunde*. Köln, Heidelberg: Mohr & Zimmer.
Schleicher, A. (1850). *Die Sprachen Europas in systematischer Übersicht*. Bonn: H.B. Koenig.
Schleicher, A. (1861/1862). *Compendium der vergleichenden Grammatik der indogermanischen Sprachen. Kurzer Abriss der indogermanischen Ursprache, des Altindischen, Altiranischen, Altgriechischen, Altitalischen, Altkeltischen, Altslawischen, Litauischen und Altdeutschen* (2 Vol.). Weimar: Böhlau.
Schlözer, A. L. (1781). *Repertorium für biblische und morgenländische Literatur* (Vol. 8). Leipzig: Wiedmann.
Schmidt, W. (1912–55). *Der Ursprung der Gottesidee. Eine historisch-kritische und positive Studie* (12 Vol.). Münster: Aschendorff.
Silvermann, S. (2005). The United States. In C. Hann (Ed.), One discipline, four ways: British, German, French, and American anthropology. The Halle lectures (pp. 255–347). Chicago/London: University of Chicago Press.
Smith, A. (1762). A dissertation on the origin of languages and of the different genius of those which are original and compounded. Considerations concerning the first formation of languages. Lectures on rhetoric and belles lettres. In J. C. Bryce (Ed.), *Glasgow Works and Correspondence of Adam Smith* (Vol. 4). Indianapolis: Liberty Fund 1985 [reprint].
Sokolovskii, S., & Tishkov, V. (1996). Ethnicity. In A. Barnard, & J. Spencer (Eds.), *Encyclopedia of social and cultural anthropology* (pp. 190–192). London, New York: Routledge.
Spencer, H. (1857). Progress: Its law and cause. In H. Spencer (Ed.), *Essays. scientific, political, speculative* (pp. 1–54). London: Woodfall & Co.
Spencer, H. (1864–1867). *Principles of biology*. Edinburgh: Williams & Norgate.
Spengler, O. (1918). Der *Untergang des Abendlandes. Umrisse einer Morphologie der Weltgeschichte.* Wien: Braumüller.
Steward, J. (1955). *Theory of culture change. The methodology of multilinear evolution*. Urbana: University of Illinois Press.
Streck, B. (2000): Kulturanthropologie. In B. Streck (Ed.), *Wörterbuch der Völkerkunde* (pp. 141–144). Wuppertal: Edition Tricker, Hammer-Verlag.
Trigger, B. G. (1998). *Sociocultural evolution: calculation and contingency*. Oxford: Blackwell.
Trubetzkoy, N. S. (1930). *Proposition 16. Über den Sprachbund. Actes du premier congres international de linguistes a la Haye du 10.15.1928* (pp. 17–18). Leiden: A.W. Sijthoff.
Tylor, E. B. (1871). *Primitive culture. Researches into the development of mythology, philosophy, religion, art and custom*. London: J. Murray.
Vacher de Lapouge, G. (1899). *L'Aryen: son rôle sociale*. Paris: Fontemoing.
Voget, F. W. (1970). Franz Boas. In C. C. Gillispie (Ed.), *Dictionary of scientific biography* (Vol. 1, pp. 207–213). New York: Charles Sribner's Sons.
White, L. (1959). *The evolution of culture*. New York: McGraw-Hill.
Wissler, C. (1926). *The relation of nature to man in Aboriginal America.* New York: Oxford University Press.
Witsen, N. (1692). *Noord en Oost Tartarye, ofte Bonding Ontwerp van eenige dier Landen en Volken, welke voormaels bekent zijn geweesst* … (2 Vol.). Amsterdam: F. Halma.
Young, T. [anonymous review] (1813). Mithridates, oder Allgemeine Sprachenkunde/Mithridates, or a General History of Languages, with the Lord's Prayer as a Specimen, in nearly five hundred Languages and Dialects, by J. C. Adelung. *The Quarterly Review,* 10(19), 250–292.
Zittel, C. (2009). *Theatrum philosophicum. Descartes und die Rolle ästhetischer Formen in der Wissenschaft.* Berlin: Akademie-Verlag.

# Darwinian Archaeology and Cultural Phylogenetics

**Daniel García Rivero**

**Abstract** This paper is a review of evolutionary thought in archaeology. It explains why and how the application of Darwinian evolutionary theory to archaeology is possible and, moreover, useful. It expounds what this scientific field gains from considering the study of material culture and, by extension, of cultural change from this perspective. After explaining the main theoretical principles, it develops a history of the application of this epistemology in archaeology, focusing particularly on the tasks of classification and sequencing of data and thus entering into the current field of cultural phylogenetics.

**Keywords** Archaeology • Epistemology • Evolution • Cultural phylogenetics • Philosophy of science • Taxonomy

## 1 Brief Notes on Philosophical Diversity in the Study of Human Cultural Change

There have been numerous epistemologies applied to the interpretation of the strictly non-genetic part of human individuals and groups, the part that we would now call cultural and which does not imply an antagonistic confrontation or incompatibility with the biological part. Archaeology is a historical science that benefits from an understanding of other fields of knowledge (such as anthropology, biology, and psychology, among others) for the purposes of a better explanation of our past and our present. In its case, the theoretical positions and paradigms of reasoning that have been developed are many and, moreover, diverse. Whereas some of these may be described as epistemologies, with theoretical bodies grounded on different fundaments, others correspond to paradigms and more aseptic and specific explicative models. Each has thrived at some particular moment or juncture and in specific

D. García Rivero
Department of Prehistory and Archeology, Faculty of Geography and History, University of Seville, Seville, Spain
e-mail: garciarivero@us.es



L. Mendoza Straffon (ed.), *Cultural Phylogenetics: Concepts and Applications in Archaeology*, Interdisciplinary Evolution Research,
DOI 10.1007/978-3-319-25928-4_3

chronological and specialist areas of our discipline. The relative acceptance and success of different theoretical perspectives have been influenced greatly by traditions and trends in the ways of doing science, if not by personal and ideological interests.

Leaving aside the issue of why there is an unequal acceptance of diverse epistemologies in archaeology, it is clear that this variability in reasoning is natural, logical, and, in every way, positive for our field of knowledge. It is natural and logical because, following philosophical principles, someone who holds that life and the world obey objective universal laws could not persuade someone who sustains that each subject has its own perception and therefore rejects the search for general laws, and vice versa. This well-known debate is exemplified by the perspectives portrayed in the philosohies of K. Popper and T.S. Kuhn. If this irreconcilable confrontation exists on such a basic reflexive level, it is not difficult to imagine how distant different postures can become at the scale of thoughts and approaches to specific historical questions. Variability is also positive in the sense that it enables to explore almost every aspect and every possibility in the analysis and explanation of natural phenomena. Indeed, the greater the degree of variability in reasoning, the higher the probabilities of obtaining the scientific keys to the validation and consolidation of knowledge.

Despite all this theoretical diversity, there is a common denominator to almost all strands of archaeological thought: the untouchable philosophical cornerstone of human intentionality as the only or, at the very best, the main mechanism driving cultural change. Archaeological theories, therefore, widely accept that history is constituted by progressive directional changes that are predetermined by people themselves. As a result, all fields of knowledge concerned with the study of human culture continue to be based mainly on an underlying Lamarckian model of change.

Intentionality is only one aspect of behavior. Moreover, it draws on our common sense, that is, on a specific culturally inherited way of understanding daily experiences and relating human actions and purposes (Dunnell 1982, referenced in Bentley et al. 2008: 115), different in each cultural setting. Intentionality, necessity and habit do not always produce change themselves. They do not usually produce intergenerational change either and, even less so, expected change. For this reason, modern science has not yet found cures for certain diseases that afflict human populations, despite the great investment of human and economic resources (intentions and purposes) in medical research. The history of technology provides many real-life examples which confirm the erroneousness of the Lamarckian model. J. Diamond (1999: 139 and ff.) compiles a number of such cases. For instance, T. Edison initially protested when his invention of the phonograph was applied to the reproduction of music on jukeboxes. Furthermore, his invention was not successful in any of the ten possible applications that he himself had intentionally foreseen. Another illustrative example of this kind is the "invention" of the motorized vehicle. Although the first internal combustion engine was built in 1866 and fitted on a four-wheeled vehicle in 1896, it was not until the First World War that it was widely applied to trucks and other means of transportation, to gradually replace the exclusive use of horse and rail transport over the following 50 years. Human intentionality is therefore a mechanism that generates variation, upon which selection may later operate (O'Brien et al. 2003: 205), as I shall explain further on. In this sense, but in relation to biology, D. Haig has claimed that epigenetic inheritance of course increases the options available to genes, but evolutionary adaptation remains the product of natural selection of random variation (Haig 2007).

A model of change based exclusively on intention has been discredited for the scientific analysis of living organisms and entities on Earth and their behavior. Considering this, it seems paradoxical that it continues to be applied uncritically to the exclusive case of *Homo sapiens*, particularly regarding behavior, but is not extended to its ancestral species. Regardless of the (natural) weight of our religious, philosophical, and cultural influences, we cannot maintain the axioms of human intentionality and the Lamarckian model if our aims are to be exclusively scientific. That is, we cannot work with the heuristics of the equation "human intentionality = estimated change," when the equation is not scientifically proven and is not always fulfilled.

Biology has scientifically validated the Darwinian model of change, which is empirically supported across diverse aspects of life. Not only is it fully supported in the fields of organic biology, but also its likelihood has been recently demonstrated by the data available in areas of knowledge concerned with the study of behavior and cultural change, particularly in certain fields of the social sciences and in archaeology (e.g., O'Brien and Lyman 2000; Shennan 2002; Mesoudi et al. 2006; García Rivero 2013). Moreover, this model fulfills the scientific values established by philosophers of science in relation to epistemology: simplicity, unifying power, fertility, internal coherence and external consistency, Popperian falsifiability, and predictive precision (McMullin 1983; Ruse 1999: 30 and ff.).

It seems then logical that this model should also be used in studying our own species. (Otherwise, it would like a present-day astronomer trying to explain the universe without taking stock of the ellipsoidal shape of our planet or without assuming the relative position of the Earth and other orbiting bodies in relation to the Sun.) The Darwinian model does not exclude the possible influence that the variation and plasticity of human behaviors may produce upon change. Rather, it sees these forces as functioning together with other mechanisms to provoke change.

## 2 The Concept of Darwinian Evolution as a Scientific Archaeological Tool

In essence, Darwinism suggests that the genetic basis of life (genotype) and all of its possible expressions (phenotype) are, together, mutable elements that change over time and space. Spontaneous generation does not exist: all expressions of life, whether somatic (physiological bodies) or extrasomatic (patterns of behavior, material culture, etc.), are a modification (to a different degree) of a previous state. But how do these modifications and changes take place?

For Darwinists, the concept of evolution is based on the notion that the variability of populations changes over time and through space. Thus, the relative frequencies of the different traits (both somatic and cultural) that constitute a population (or a system comprised by several populations) do not remain constant, but rather vary in time. The forces that drive change do not reside exclusively inside the organisms, as may have been suggested by the pre-Darwinian concept of transformism. Apart from the internal variation of organisms, produced by genetic mutation and behavioral innovation, for Darwinists, an important means of change is found outside of the organisms. Although the existence of plural mechanisms has been largely debated –even by Darwin himself (1859: 206) – the definitive cause of Darwinian change is the process known as natural selection.

Selection refers to any situation in which some type of pressure causes a differential reproductive bias in the traits of a population (or of the populations in a system). This pressure (generally qualified as "selective") creates a funnel or bottleneck that prevents the replication of all the variations, thus favoring higher rates of replication and offspring of just some of those variations. Such selective pressures may be generated by of a group of organisms or of more general environmental variables and may affect a single population or a large number of them. For this reason, it may be said that Darwinian change is external, since selective pressures act sporadically and in accord with the interactions between the elements in a system.

Selection therefore constitutes the main mechanism due to which the relative frequencies of traits and organisms are modified over time. The traits or organisms that, for whatever reason, fail to reproduce or replicate will not leave descendants or copies of themselves. When a group of traits or organisms does not reproduce sufficiently (the threshold varies on the contextual circumstances), it will become extinct, while a different group of traits or organisms, which in that particular setting display greater differential reproduction, will take its place.

In addition, the process of selection encompasses the variables of time and geographical space, overlying a multitude of factors such as mutation, drift, migration, isolation, and countless possibilities of relationships and levels of association between organisms and populations. The result is speciation: the emergence of new species from preexisting lineages.

The features that are required for a system to be analyzed from a Darwinist perspective are: (1) variation, (2) differential reproduction of variation, and (3) inheritance, or the genetic or cultural replication of inherited traits. Figure 1 illustrates this basic model.

Whereas this scheme of change has long been established in biology with the same terms, anthropologists and archaeologists have also used evolutionary notions for a long time, but not until very recently (see below) have they really worked with this model. In biology, the term genotype is applied to the DNA that synthesizes the molecular information expressed in the phenotypic traits (eye color, hair color, etc.). In the case of culture, the genotype should be understood as all of the information that is stored and culturally transmitted – by various mechanisms – between human minds, for instance, the processes of pottery production that are codified in the potter's mind. In this case, the information is expressed phenotypically in the pottery vessels, upon which selection operates. Thus, behaviors and material culture are phenotypes expressed through the minds of individuals and function in a similar way to that of the genetic phenotype.

The translation of the model of genetic transmission to the study of cultural transmission does not imply, firstly, that the transmission of both types of information (genetic and cultural) necessarily occurs independently, since both effectively belong to a single integrated system: an organism or, at a higher level, a population. In this sense, the "cultural genotype" may be considered as the sum of simultaneous cultural and genetic information, materialized in physical bodies, their behaviors, and the material culture that they produce (Bentley et al. 2008: 114). Secondly, the application of this model to the study of cultural information does not imply that the structure and transmission of cultural information take place in the exact same ways as in the case of genetic information. However, given that this model has enabled great advances in the understanding of the organic world, it would be useful to establish an initial scheme to begin to study how cultural information may be structured, stored, and transmitted.

The sum of the cultural genotypes and phenotypes of a society includes a very wide range of traits: ideas, beliefs, values, behavioral patterns, languages, extremely varied material objects and utensils, etc. These cultural traits are not constant over time, or throughout space, between different groups and societies. The diversity displayed by the anthropological and archaeological records is the product of the evolution of cultural traits over time and space, by means of diverse mechanisms including variation, inheritance, and processes of bias such as natural selection. In our analyses, if we think in terms of populations and we create cultural units, similar to phonemes and morphemes in linguistics, measurable and appropriate for our hypotheses and methodological tools, then we will be able to see how such cultural variants are distributed within and between populations across different periods. Different relative frequencies will enable us to formulate and test hypotheses – essentially inspired by Darwinian principles – on, for instance, the differential reproduction of cultural traits in populations through time. Such an approach will help to understand the reasons behind the proliferation and decline of different archaeological materials. In other words, it will provide explanations of cultural change.

## 3 The Darwinian Study of Cultural Traits and Systems

The study of the evolution of cultural traits and systems requires such units to be measured and tracked before they are analyzed and explained following the Darwinian principles of change. The search for historical realities, or for historical truths in the Popperian sense, relies on the orderly development of questions and answers, on a systematic framework that enables the integration of our conjectures, hypotheses, inferences, and comparisons, thus creating a cumulative and collective process of knowledge.

Among the earliest forms of scientific reasoning are the methods based on analogy, which can be traced back at least to Aristotle (cf. Arist. *Topica* I, 17–18).

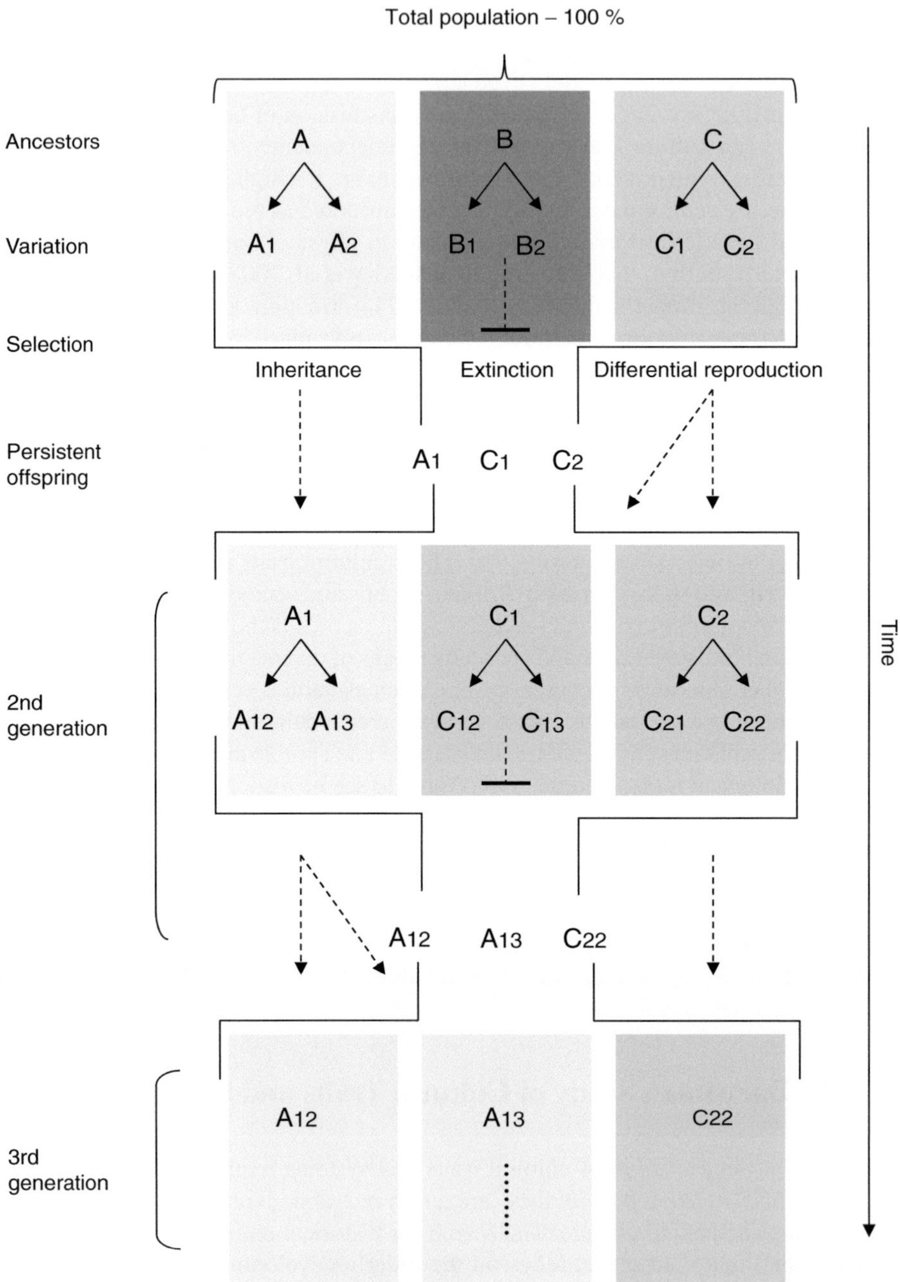
Total population – 100 %
Ancestors
Variation
Selection
Persistent offspring
2nd generation
3rd generation
A
B
C
A1 A2
B1 B2
C1 C2
Inheritance
Extinction
Differential reproduction
A1 C1 C2
A1
C1
C2
A12 A13
C12 C13
C21 C22
A12 A13 C22
A12
A13
C22
Time

Indeed, the comparison of two parts, quantities or assemblages, enables inferences to be made about their common aspects. Today it is difficult to imagine a science, including archaeology, without the comparative method. In historicist archaeological literature, for instance, it is usual to find frequent references to objects that are seen as more or less similar to those under study. In such cases, the method followed involves the comparison of elements from the known material record in the search for similar archaeological objects or assemblages.

Although interesting, the application of a comparative method based solely on analogy poses a serious problem for the task of classification, which has barely been addressed by archaeologists: the comparative method does not enable the assessment of the nature of the observed similarities between different objects or assemblages. In other words, the consideration of whether the analogies are due to kinship, parallelism, evolutionary convergence, or horizontal transmission is overlooked (what is known as "Galton's problem").

Kinship refers to when two objects or assemblages are related to a common ancestor. For instance, in the Phoenician diaspora of the early first millennium BC throughout the western Mediterranean, two populations, unconnected since they departed from the homeland at different times and setting different destinations,

**Fig. 1** In the upper part of the figure, the letters A, B, and C represent the three first ancestors of the example. They could correspond to three different organisms, three different patterns of behavior, three pottery types, or three possible cultural traits or variants existing in a given population and generation. In order for the example to be more simple and illustrative, the size of the population will be constant (there will only be three niches or positions in this ecosystem). Time is represented vertically, from top to bottom. The different stages of the process are indicated for the first generation. The traits A, B, and C must be understood as the three cultural variants existing in the first generation which, under the effect of selective pressures or a bottleneck, will compete over their replication in the second generation. The three initial traits in this example are closely related (for instance, pottery types) and are originally present in equal frequencies: each corresponds to a third of the population. All three produce two new traits through modification (A produces traits A1 and A2). However, not all six of these new traits can be reproduced in the following generation because demand creates pressure and competition between the traits. Darwin used the expression "*struggle for life*." In the second generation of our example, the descendants are three traits (A1, C1, and C2). In this hypothetical case, the two variants derived from the ancestor B have become extinct. The third of the population occupied by B has been replaced by the traits derived from ancestor C (C12 and C13). This increase of C in the relative frequency of the population over the generations (already visible after a single generation) is known as "differential reproduction" (the *dark gray shade* of the first generation becomes extinct in favor of the medium *gray shade*). After just one generation, the initial variants have become modified, have evolved, and have even been replaced by others. However, we may draw vertical (broken) lines that show the strong genealogical relationships between ancestors and descendants. These illustrate their degree of *fitness* or adaptability which, to a great extent, influences succession. Effectively, the transmission of information from one generation to the next takes place by means of diverse mechanisms of inheritance. If we continue down the figure, in time, we find the same processes in another generation and so forth

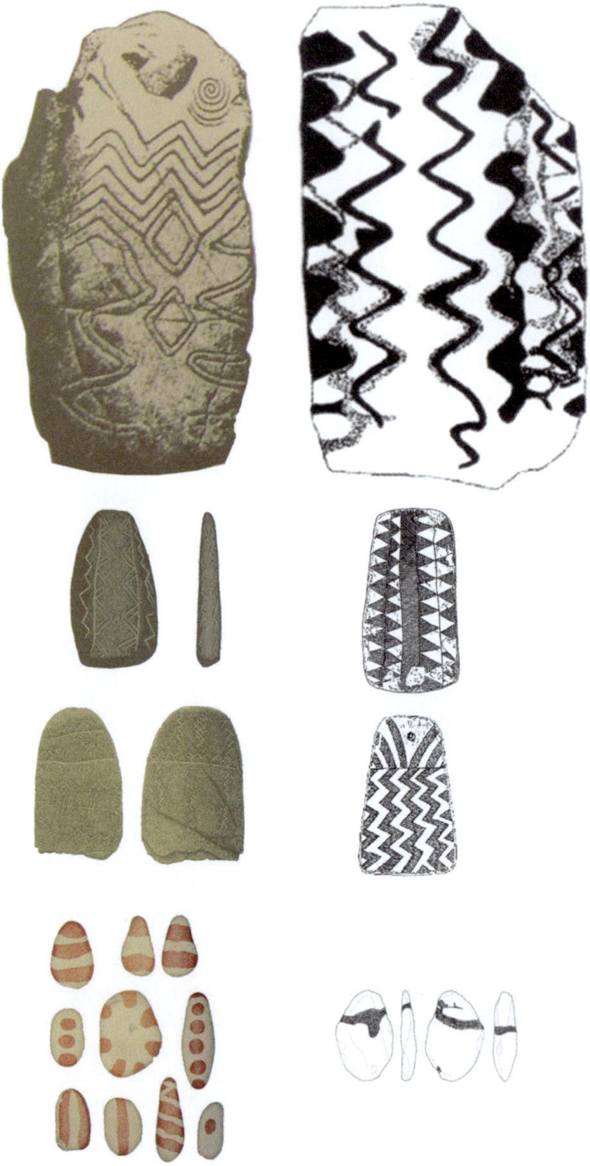

**Fig. 2** This figure visually compares elements of the archaeological records of Native Americans (left-hand column) and the late prehistory of the Iberian Peninsula (right-hand column). The decoration of Iberian megalithic orthostats and engraved slated plaques is apparently very similar, almost identical, to that created on spatial and territorial markers and on portable artifacts by Native American societies. We may also observe strong similarities between North American painted pebbles and the same artifact type from the Cantabrian Azilian (Northern Spain). This is an example of parallelism or evolutionary convergence between two populations and cultures that are completely separate, both geographically and chronologically (Figure elaborated from Breuil (1935), Carpenter and Schuster (1988) and Lillios (2004))

may share and display a common knowledge of the iron production techniques that existed in the original population from which both descended.

Parallelism, on the other hand, refers to a trait that appears in two different places and times, without there being a phylogenetic connection between them, that is, without any common ancestor. Figure 2 compares particular elements of the American archaeological record to some from the Iberian Peninsula's prehistory. The similarities are due to a chance phenomenon that has caused material parallelism between two populations, geographically and chronologically separated. In this example, the spatial and temporal disconnection of these two settings can be assumed. But how would we proceed if we were dealing with archaeological materials from the same continent and chronology? It is practically impossible, or at least ambiguous, to establish whether similar traits are due to parallelism or kinship based on the observation of similarity alone. This is the main shortcoming of methods based on the analogy of resemblances and similarities, since these are insufficient and incorrect for the determination of groups and classifications.

Evolutionary convergence refers to a trait found in two different places and times, with no phylogenetic connection, but in which the emergence of the trait is due to similar processes of adaptation, caused by similar selective pressures in different contexts. An illustrative example of evolutionary convergence is the "feather or fur color" character in the following case. The rock ptarmigan and the snowy owl, as well as the polar bear, have white feathers or fur, as a result of the selective pressures that exist in snowy environments. The external color of the organisms of these species has been modeled by natural selection in populations that arrived and, over generations, thrived in arctic regions. This example shows the emergence or modification of traits toward identical states in different populations that do not share a recent common history. It is probable, for instance, that the emergence of agricultural societies in different primeval locations throughout the world may be due to evolutionary convergence under similar selective pressures and processes.

Finally, another cause of similarity between two populations or assemblages is horizontal transmission, which takes place between two interacting populations that transmit the trait in question. The examples of horizontal transmission of cultural traits are innumerable: Roman populations transferred many cultural and material traits to diverse ethnic groups with whom they interacted in widely variable geographical regions. By a similar process, many present-day societies all over the globe have adopted cultural trends from the United States of America, from food and material commodities to television formats of artistic expression.

As I will argue below, phylogenetics is the only method of classification that enables us to confront and analyze this issue in detail and therefore to put forward more rigorous and consistent classifications and models of historical sequence.

Sequencing is another fundamental method of scientific reasoning in the historical sciences. The purpose of this method is to place different taxa and assemblages in relation to each other within a same historical process. This method has also been used in science for a long time. As I shall discuss in section 5 of this chapter, anthropologists and archaeologists have used this method since the mid-nineteenth century and modern biology and history would be inconceivable without the multiple techniques of sequencing. However, in this case also, the theoretical nature of sequencing has barely been addressed in archaeology. Generally, historical evolution has been accepted as solely linear, involving the transition of cultures through different states along a single straight path (phyletic sequence). This vision had even been applied to the development of our ancestral hominid history until a few decades ago. Nowadays, the historical understanding of evolution is very different and is based on graphic tree representations of a divergent model of life in which a series of branches emerge from a single trunk. Indeed, rather than aligning all change on a single line of descent (as is the case in phyletic models), phylogenetics acknowledges the divergent nature of evolution, in the shape of a tree.

Another important aspect of scientific reasoning is the experimental method, that is, the testing of inferences and hypotheses. The scientific method can be traced back to Galileo, while its modern standard procedure was detailed philosophically by Karl Popper. Although we often relate experimental science to the exact sciences (mathematics, physics, chemistry, etc.), it is possible to carry out experimentation (in the sense of testing) in the fields of historical science (anthropology, archaeology, biology, paleontology, etc.) (e.g., Clark and Stafford 1982; Aldenderfer 2005; McGlade 2005). However, because part of human behavior may well be extinct, the questions and tests must be indirect (based on the available record). Of course we may formulate as many inferences as we wish, but we need to test our hypotheses in order to clarify any doubts surrounding their validity. Archaeological methods and interpretations often lack such testing, due to the perceived absence of means of direct (internal) testing. Evolutionary methods and phylogenetics, particularly, enable us to test sequences of classification and the underlying historical hypotheses using a statistical numerical basis in accord with the principle of refutation and the systematization and accumulation of knowledge.

To end this section, I now formulate clearly the advantages of implementing an evolutionary systematic archaeology. This view enables us to align congruently, in a phylogenetic model, a theoretical postulate (evolution by descent with modification) with the variation displayed by the archaeological record. The systematization of this alignment allows for the ordered and progressive development of knowledge by means of tested inferences and deductions. (Darwinian) Evolutionary theory and systematic phylogenetics are compatible and congruent with each other and also with the premise of the objective search for historical reality of truth in the Popperian sense. As it did in biology several decades ago (O'Hara 1997), the conjunction of systematic phylogenetics and population thinking noted above may constitute today the research tool of greatest potential in the field of archaeology.

## 4 Phylogenetic Taxonomy

The vast majority of classifications found in archaeological studies belong to the taxonomic school of "phenetics." These classifications are based solely on the criterion of resemblance or similarity between specimens. The specimens under study are sorted into inclusive nested groups on different levels. Such hierarchical classification schemes follow the model established in 1735 by C. Linnaeus in his *Systema Naturae*. This cornerstone of taxonomic studies is still in use today, although in the fields of linguistics and biology it was complemented by the concept of evolution over a century and a half ago, making it more fertile and operative and more scientifically correct. Linnaeus believed that the species and groups of organisms that he classified were completely static, immutable elements created by God in permanent form, from the beginning until the end of time. Once J. B. Lamarck (1809) and Ch. R. Darwin (1859), among others, demonstrated that organisms change over time, it became necessary for this taxonomic model to consider a new element of dynamism:

transformism for Lamarck and evolution by natural selection for Darwin. What is particularly interesting for us to note here is that, from that point onward, classifications could no longer be based solely on the general resemblance of organisms, but would require an additional dimension able to account for the information transferred (transmitted) between such organisms. In biology, from then on, the notions of inheritance and kinship gained importance, and the first graphic and conceptual sketch of the tree of life was made following the same structure as the previous (immutable) model.

It was not until the mid-twentieth century that this taxonomic conception was consolidated and methodologically formalized by W. Hennig (1950, 1965, 1966), after important advances in the knowledge of the genetic support for life (the information stored in the genes) had been made. This new taxonomic school became known as "systematic phylogenetics" and is nowadays known as "cladistics" (cladistics and evolutionary taxonomy, another school that I shall define below, constitute the field generically called "phylogenetics" or phylogenetic methodology). Despite its central role in our discipline, most archaeological taxonomy practiced today, unfortunately, appears not to have moved beyond the phenetic school and the Linnaean immutable model, overlooking the implications that the evolutionary principles have on material culture as well. The weakness of phenetics, as mentioned above, lies on its inability to address the different possible reasons for the resemblance of traits between organisms or objects. Moreover, phenetics lacks a theoretical model and thus constitutes an empty taxonomic method that struggles to tell a story beyond the classification models themselves. In cladistics, in contrast, Darwinian evolution provides a theoretical framework for the Linnaean taxonomic system, where system and theory become interconnected (cf. O'Brien and Lyman 2003: 112).[1]

The only taxonomic methodology that currently enables us to study transmission is phylogenetics. Phylogenetic methodologies are based on transmission, whether genetic or cultural. Phylogenetic studies of genetic and somatic information are quite common and can be found in approaches to subjects close to our field of research, for instance, in classificatory proposals of hominid evolution. These studies generally use morphometric information provided by paleontological remains or genetic information extracted from genomic sequences. Phylogenetic approaches to culture are based on extrasomatic traits, for instance, particular cultural patterns and material culture itself. Whereas genetic transmission takes place through the replication and transference of the information contained in the genome, cultural transmission operates by means of diverse mechanisms: observation, emulation, imitation, conditioned stimulus, direct teaching, spoken language, writing, etc., within a range of processes of variable complexity (Richerson and Boyd 2005: 63).

In its basic form and in order to establish groups, phylogenetics uses the information transmitted according to the laws of kinship, that is, vertically transmitted information, for instance, from parents to children, from master potters to apprentices, from one generation to the next within the same population, etc. (Tehrani and

[1] I will not go into further detail on this question since that would take us away from the central theme of this contribution; however I refer the reader to a synthesis of the main theoretical principles of classification and the three existing taxonomic schools (García Rivero 2010b).

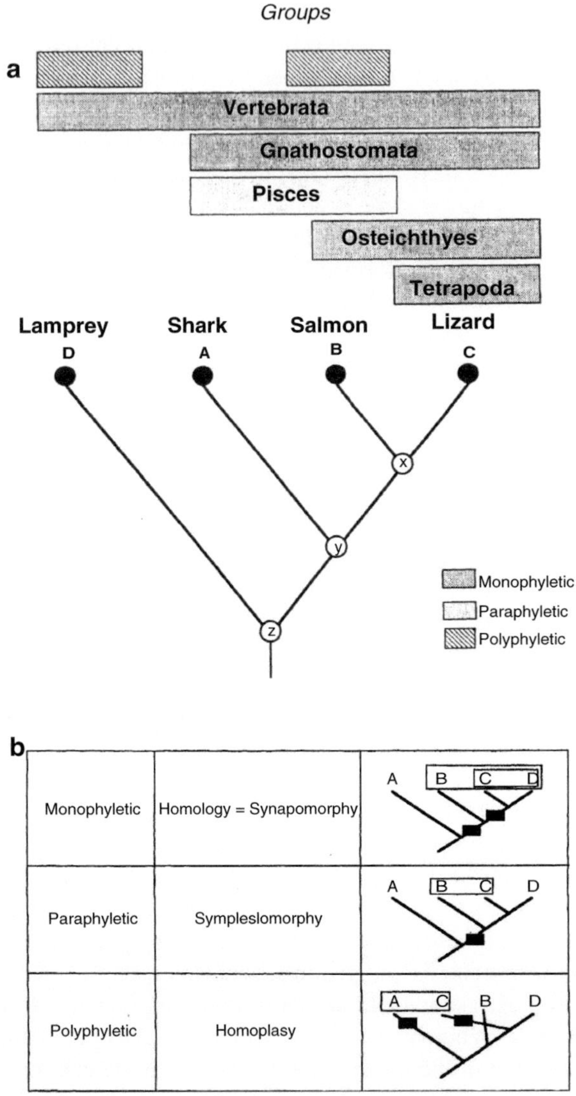


**Fig. 3** (**a**) Types of taxonomic groups; (**b**) types of characters accepted for their construction. Phenetics orders taxa based on polyphyletic groups, cladistics on monophyletic groups, and evolutionary taxonomy on paraphyletic groups (Figure elaborated from Kitching et al. (1998, Figs. 1.8 and 1.10))

Riede 2008). Cladistics excludes analogue characters, those that display a similar state in different species without evidence of a common genealogy between them. Instead, it uses homologue characters, particularly those called "shared derived characters" or "synapomorphies" in the phylogenetic argot (Fig. 3). These correspond to traits that, for the purpose of establishing groups, are shared by all sibling taxa and their common ancestor, but not with the ancestor immediately prior to the

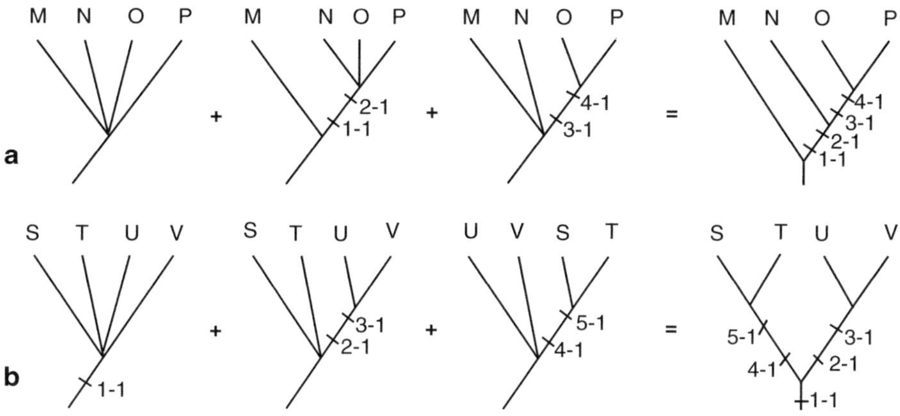


**Fig. 4** (**a**, **b**) Two examples of the rule of inclusion/exclusion – or Hennigian argumentation – combining the information of different series of transformation of states of three provisional representations. The solutions of both series are the two phylogenetic trees on the right (Figure elaborated from Wiley et al. (1991, Fig. 2.2))

latter (where the trait to be shared with this ancestor as well would be called a "shared ancestral character"). Synapomorphies constitute what are technically known as "monophyletic clades." This unit is the only criterion considered in cladistics for the ordering of taxa under the Darwinian taxonomic model.

"Hennigian argumentation" or "the rule of inclusion/exclusion" is the basic procedure for the construction of a cladogram (a diagram of branches that represents a hypothesis of hierarchical relationships between taxa on the basis of the shared derived characters). A single graphic representation will thus include the available information for all of the series of transformations of the considered characters or traits (Fig. 4).

Within what is generally known as phylogenetics, as well as cladistics, there is another taxonomic school called "evolutionary taxonomy." It appeared soon after cladistics, although both developed in parallel. The pioneering studies in this new taxonomic perspective belong to Ernst Mayr (1969). While phenetics uses only the criterion of overall similarity to order taxa, and cladistics is based exclusively on phylogenetic kinship, evolutionary taxonomy accepts both criteria – similarity and genealogy – as useful elements in taxonomic research.

I shall not offer an explicit comparison of these two phylogenetic schools of taxonomy – cladistics and evolutionary taxonomy – since this question has been dealt with elsewhere (García Rivero 2010b) and since both correspond to different perceptions and research processes motivated by and leading to different aims. While evolutionary taxonomy contemplates model and process simultaneously, searching for and analyzing genealogical elements in ecological niches, cladistics is concerned solely with the model, accepting Darwinian theory as the backdrop in which all processes are proposed and explained (O'Brien and Lyman 2003: 96).

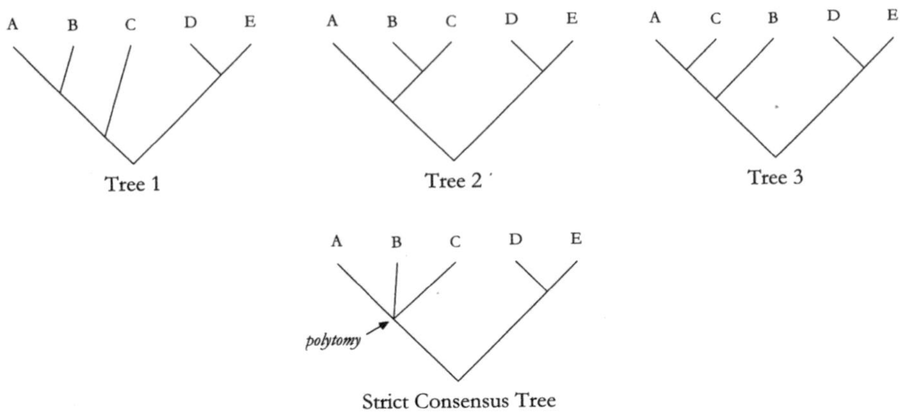


**Fig. 5** Hypothetical situation in which there are three possible cladograms of equal parsimony but which display incompatible options for the clade constituted by specimens A, B, and C. In this case, a solution of strict consensus is sought (there are different types of consensus), which combines all of the information from the three upper cladograms into the lower tree (After O'Brien and Lyman (2003, Fig. 3.9))

However, I must note that cladistic methods are based entirely on the "principle of parsimony" as the criterion of selection between different possible phylogenetic hypotheses, while evolutionary taxonomy also makes use of other statistical principles such as "maximum likelihood." In actual case studies, one often obtains more than one classificatory proposal. Diverse graphic representations representing different hypotheses of how the taxa under study are related to one another may be obtained, according to the selected traits. Parsimony is the criterion adopted by cladistics to select among the different trees; this is justified by the assumption that nature favors simplicity. When facing various possible explanations of a process or phenomenon, the most simple and easy solution will be retained as the most probable (Mayr and Ashlock 1991: 216; Goloboff 2003). When facing different cladistic hypotheses, the one that requires the least evolutionary changes will be retained. If there were several cladograms of equal parsimony (different representations of a single problem with the same number of evolutionary changes), it would be necessary to create a consensus tree (Fig. 5). In evolutionary taxonomy, as mentioned above, other principles are also used to choose and discern between different graphic representations. Maximum likelihood and Bayesian theorem of probability are currently the most common (e.g., Lewis 2001; Holder and Lewis 2003; Ronquist 2004). Such methods require an explicit model of the evolution of the characters or parameters. The likelihood of each state in every position is estimated, revealing the probability of the data according to the model. The phylogenetic tree with the highest rate of likelihood will then be selected as the most probable.

Before concluding this section, we must consider a final technical question that concerns all phylogenetic approaches, whether based on cladistics or evolutionary taxonomy. As noted above, phylogenetic studies are based on vertically transmitted information, according to the laws of kinship or, in other words, on the information that can be explained by an ancestor-descendant model. Often, in real natural phenomena, homologue characters do not account for all of the information transmitted between taxa and populations. Usually, there are traits which have not strictly been transmitted along the vertical axis, thus blurring the phylogenetic signal that might be reconstructed by the methods outlined above. This technical obstacle receives the name "homoplasy" and may be caused by several circumstances: (1) parallelism or evolutionary convergence, or the independent emergence of the same trait in unrelated taxa; (2) horizontal transference, or the transmission of a trait between coexisting taxa or populations; and (3) the reversion of states in characters, that is, the spontaneous reversion of a trait to a previous or ancestral state.

Several critical works addressing the controversies of using of phylogenetic methods to study of cultural evolution have been published, particularly in archaeology (e.g., Gould 1987: 70; Moore 1994; Terrell and Steward 1996; Tëmkin and Eldredge 2007; Schiffer 2008). One of the most common and serious objections is that related to the problem of homoplasy, and it is often underlined that cultural evolution is different in nature to organic evolution, since the former is highly reticulated (cf. O'Brien et al. 2008: 48). The discussion of the dichotomy between both models is not new (e.g., Bellwood 1996) but has recently led some researchers to assess the importance of *phylogenesis* over *ethnogenesis* in real anthropological and archaeological case studies (e.g., Collard and Tehrani 2005; Collard et al. 2006).

Regardless, it is worth noting that this suggested problem is not exclusive to the application of phylogenetics to culture (including material culture). Molecular biologists and researchers involved in the study of microorganisms and plants deal with this issue on a daily basis. The importance of horizontal transmission between genomes and bacteria has been recognized (Margulis and Sagan 2002), and high rates of horizontal transmission between many species and families of plants and animals have also been recorded. As noted by M. J. O'Brien and R. L. Lyman – with reference to relevant literature (cf. 2003: 104–105) – even hybridization appears to be well documented in biological evolution, particularly among plants, where it may account for up to 20 % of variation, especially in angiosperms. In the animal kingdom, albeit with lower rates, the case of birds may be illustrative (Laskowski and Fitch 1989).

Some studies have assessed the performance of taxonomic techniques by addressing the impact of hybrids and the influence of homoplasy on the indexes and coefficients currently used in cladistics (Sanderson and Donoghue 1989; McDade 1992; Baroni et al. 2004, 2006; Greenhill et al. 2009; Currie et al. 2010; Muscio 2010). These studies have demonstrated that, although the effects of homoplasy are not in fact so pronounced, some weaknesses do exist which deserve further study in order to minimize such obstacles. However, the observation of horizontal transmission between populations of microorganisms, and plant and animal species, has not led to the rejection of phylogenetic analyses in biology (cf. O'Brien et al. 2008: 48). Indeed, how would biology function without phylogenetic methods? Despite awareness of rates of horizontal transmission in genetic information (Woese 2004), and of the existence of hybrids, biologists do not disregard these methods because they recognize that these allow for more reliable and scientific taxonomic reconstructions than any other method currently available.

The matter of horizontal transmission and homoplasy may thus be considered a methodological problem, rather than a theoretical issue (Bellwood 1996), that can be tackled by the study and refinement of the phylogenetic techniques themselves. J. S. Farris already suggested that the processes of hybridization had been underestimated, several decades ago. He attempted to make up for this problem through the computation of Wagner trees and the construction of networks (Farris 1970). More recently, some authors have presented alternative models to counteract the effects of homoplasy and horizontal transmission, as well as the existence of polytomies (an internode – a hypothetical ancestral taxon – on a cladogram that groups together more than two taxa [cf. Fig. 5]). Techniques based on networks are being developed for this purpose, using diverse methods such as *split decomposition* and *neighbor net* (Bryant and Moulton 2002; Makarenkov and Legendre 2004; Bryant et al. 2005), "reconciliation analysis," and "mixed Bayesian models" (Gray et al. 2007) and the so-called rings of life (Rivera and Lake 2004).

## 5 Cultural Phylogenetic Studies

In this section I offer a summary of studies that embrace an evolutionary conception of cultural change and, particularly, of recent work which applies systematic methods and techniques toward its reconstruction. Although I shall refer to some pioneering applications of these methodologies in various fields of social science, such as linguistics and anthropology, the main aim is to illustrate the history of their development in archaeology, specifically.

Even before naturalists, some linguists had perceived that historical languages must have had their origins in ancestral forms and therefore must have been subject to some form of modified evolution, in the sense later described by Lamarck and Darwin. J. J. Scaliger (1540–1609) realized that some characters were inherited between languages and recognized their importance to reconstruct language relationships (Atkinson and Gray 2005: 515). M. Z. van Boxhorn suspected that the similarities between Indo-European languages might have been due to the existence of a primitive common language he named "Scythian." Not limiting himself to a lexical comparison between individual languages, he compared whole languages "as organic systems of grammatical regularities," which led him to recognize the Indo-European family in 1643 (van Driem 2001: 159).

Also noteworthy is the work of W. Jones (1786) who continuing in this line of enquiry suggested that the great similarity between Sanskrit and European languages, such as Latin and Greek, could not be coincidental. He proposed that these had emerged from a single ancestral language, not so remote in time, now known as Proto-Indo-European. The speakers of this language would have spread across Eurasia, and different isolated populations would have shaped different linguistic communities and, with the passing of generations, different languages.

Very soon after the publication of *The Origin of Species*, a genealogical model for Indo-European languages was put forward by A. Schleicher (1863). As is evident

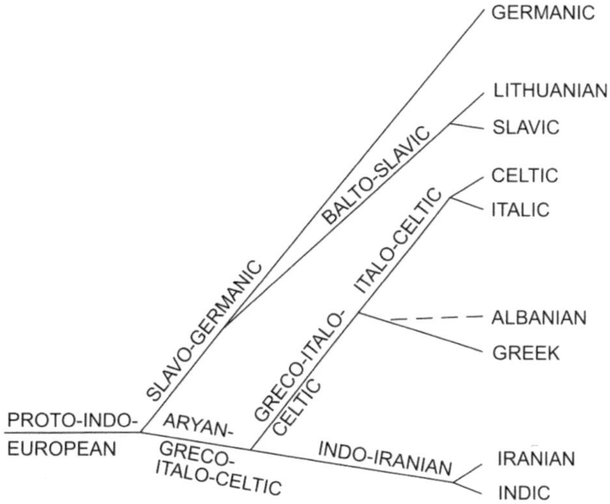


**Fig. 6** Phylogenetic tree (based on kinship) as a model of the evolution of Indo-European languages, after Schleicher (1863, taken from Forster and Renfrew 2006, Fig. 1.2)

from the title, Schleicher's work was strongly influenced by Darwin's fundamental theory. Therefore, it is not surprising to find that the classification of the evolution of some Indo-European languages strongly resembles present-day graphic representations in phylogenetics (Fig. 6).

In the field of linguistics, the evolutionary conception was not constrained to the study of the evolution of spoken languages. As early as the first half of the nineteenth century, scholars and analysts of manuscripts such as K. Lachmann and C. Gottlieb Zumpt claimed that the systematic study of variants (changing elements) in script traditions is genealogical in nature. Manuscript traditions have been addressed more recently in several works (cf. Spencer et al. 2006: 67; Lipo et al. 2006b: 5).

During the twentieth century, and particularly in recent decades, linguistic studies based on truly phylogenetic perspectives and methods have also been developed. Some have been based almost purely on cladistics whereas others have combined the simultaneous use of inheritance and distance (e.g., Platnick and Cameron 1977; Renfrew 1992; Gray and Jordan 2000; Holden 2006; Gray and Atkinson 2003; Rexová et al. 2003; Greenhill and Gray 2005; Holden et al. 2005; Bryant et al. 2005; Forster and Renfrew 2006; Atkinson and Gray 2006; Nicholls and Gray 2008; Jordan and O'Neill 2010).

In the nineteenth century, North American and European scholars from the fields of anthropology and archaeology developed pioneering studies of cultural evolution. These, however, were based on unilinear and progressive models that viewed cultural change as a directed process from primitive forms (savage and barbaric societies) toward more perfect and elaborate states (civilizations), in which human necessity was put forward as the driver of evolution. The works of E. B. Tylor (1871) and L. H. Morgan (1877) are well-known examples of these early theories.

The first 60 years of the twentieth century saw the development of new models of cultural evolution based on the Darwinian idea of descent with modification

(Steward 1955; White 1959; Service 1962; and cf. O'Brien and Lyman 2003: 3). Nonetheless, the perceived driving force of change remained basically the same as in previous models (human necessity and intentionality). The causes and processes that produced cultural traditions and change were not addressed analytically, and it was generally assumed that independent invention or cultural borrowing (diffusion) accounted for most cases (cf. O'Brien and Lyman 2003: 4).

In this first phase of cultural evolution studies, the techniques and methods applied, specifically to the evolution of material culture, aimed at creating classifications or lineages that would show the relationships between archaeological objects over time. For example, "seriation" was founded on the assumption of historic continuity and some sense of inheritance and enabled the construction of chronologically ordered series of materials, under the assumption that the degree of similarity between two objects was related directly to their temporal proximity.

M. J. O'Brien and R. L. Lyman have, in some of their publications (O'Brien and Lyman 1999, 2003; Lyman and O'Brien 2006), highlighted the main landmarks in the historiography of this technique. The studies by J. Evans in the mid-nineteenth century on British Protohistoric and Roman coins are probably one of the earliest examples of seriation (Fig. 7: 3a). A. H. Pitt Rivers also published in 1875 a series of archaeological and ethnographic objects belonging to Melanesian cultures, including rowing oars from New Ireland in the Bismarck Archipelago (Fig. 7: 1a). In 1899, W. M. F. Petrie classified the artifact assemblages recovered from ancient Egyptian tombs, focusing in this case on pottery vessels (Fig. 7: 3b). In 1915, B. Dean studied the evolution of medieval and modern metal helmets (Fig. 7, 2a) and created other divergent seriations of different types of swords, although these were not published at the time (cf. Lyman and O'Brien 2006). In 1917, A. V. Kidder also applied a seriation technique to the classification of the pottery of Pecos Pueblo in New Mexico (Fig. 7: 1b). Some years later, in 1937, E. B. Sayles would order the elements known as "manos" and "metates" (pestle and mortars) from Snaketown, Arizona (Fig. 7: 2b). Also in 1937, H. S. Colton and L. L. Hargrave (1937) created a classification of the pottery series and types from the southwest of the United States (Fig. 7: 2c).

**Fig. 7** (**1**) Examples of archaeological sequences based on the anagenetic model of evolution: (*1a*) seriation of oars from the Bismarck Archipelago, New Guinea, by Pitt Rivers in 1875; (*1b*) seriation of the pottery decoration motifs of Pecos Pueblo, New Mexico, by Kidder in 1917; (*1c*) seriation of artifacts, by J.A. Ford in 1962. (**2**) Examples of archaeological sequences based on the cladogenetic model: (*2a*) genealogical evolution of medieval and modern metal helmets, by Dean in 1915; (*2b*) proposed development of pestle and mortars from Snaketown, Arizona, by Sayles in 1937; (*2c*) hypothetical representation of the relationships between pottery series (*capital letters*) and their types (*lowercase*), by Colton and Hargrave in 1937. (**3**) Examples of archaeological sequences based on the evolutionary models of anagenesis, cladogenesis, and reticulation: (*3a*) seriation of British Roman coins, by Evans in 1850; (*3b*) genealogy of pottery in predynastic Egyptian tombs, by Petrie in 1899 (Figure elaborated from O'Brien and Lyman (2000, Fig. 6.6; 2003, Figs. 1.1, and 1.5), Lipo et al. (2006b, Fig. 1.2), and Lyman and O'Brien (2006, Figs. 5.1, 5.3, 5.6, and 5.10))

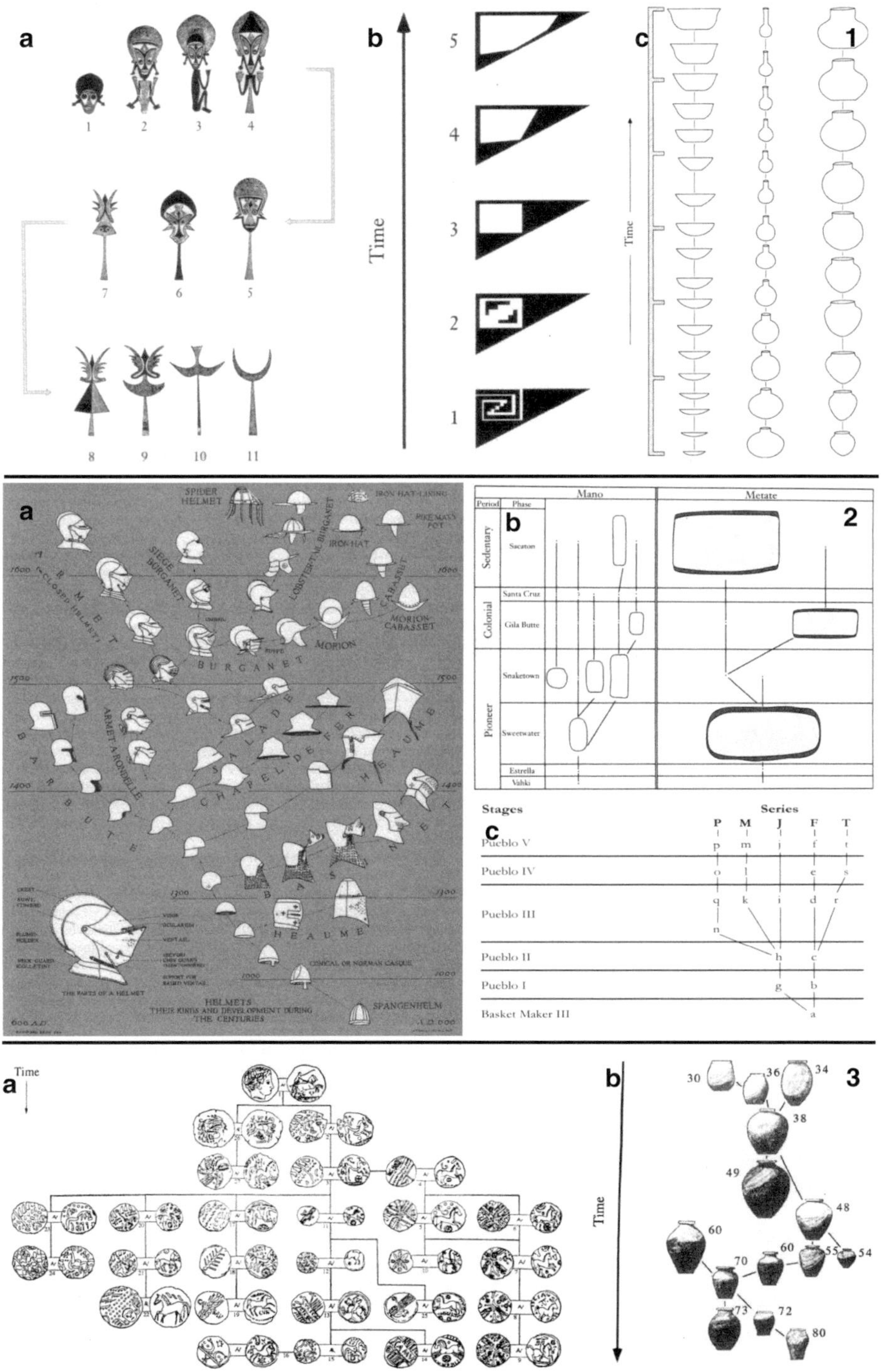

a
b
c
1
Time
2
Mano
Metate
Period
Phase
Sedentary
Sacaton
Santa Cruz
Colonial
Gila Butte
Snaketown
Pioneer
Sweetwater
Estrella
Vahki
SPIDER HELMET
IRON HAT
MORION
MORION-CABASSET
BURGANET
SALADE
CHAPEL DE FER
HEAUME
ARMET A RONDELLE
BARBUTE
BASSINET
SPANGENHELM
CONICAL OR NORMAN CASQUE
THE PARTS OF A HELMET
HELMETS THEIR KINDS AND DEVELOPMENT DURING THE CENTURIES
Stages
Series
P M J F T
Pueblo V
Pueblo IV
Pueblo III
Pueblo II
Pueblo I
Basket Maker III
3

It is however necessary to distinguish among the several types of seriation in the works mentioned above, which display different conceptions of classification according to the evolutionary model which was applied (anagenesis, cladogenesis, or reticulation). The first of these, based on the concept of anagenesis, perceives change as gradual and unilinear: different taxa succeed and substitute each other along a single vertical axis (O'Brien and Lyman 1999). Some of the studies cited above (Fig. 7: 1) exclusively used this type of seriation, also known as phyletic, for example, the works by Pitt Rivers and Kidder and the 1962 theoretical discussions of A. Ford (cf. O'Brien and Lyman 2003: 7). The second, based on cladogenesis, was applied by Dean, Sayles, and Colton and Hargrave (Fig. 7: 2). It is based on the idea that an ancestral taxon gives way to two new taxa and is essentially the model that would be followed by systematic phylogenetics or cladistics. Dean, for instance, combined the cladogenetic and anagenetic models (Fig. 7: 2a). The third model, reticulation, is based on the idea that two taxa may hybridize to the point of creating a new third taxon. This model was used, along with the two others, by Evan and Petrie (Fig. 7: 3).

There are also some variants of seriation that reflect another class of information, not pertaining to the individual development of the taxa but to the archaeological assemblages. A. L. Kroeber's 1916 work on the pottery assemblages of the Zuni people of New Mexico is the first example of "frequency seriation" (cf. O'Brien and Lyman 2003: 11–12). "Occurrence seriation" was introduced later, in the mid-twentieth century (cf. Lyman and O'Brien 2006: 71). While frequency seriation is based on the relative quantification of the supposed historical types, usually expressed as a percentage, occurrence seriation considers only the presence and absence of types. Both variants assume, as the previous models did, the inference of chronological information based on historical continuity. Later, numerous seriation techniques based, for example, on multivariate methods, have been developed (e.g., Doran and Hodson 1975; Fernández Martínez 1985; Madsen 1988).

This brief summary highlights the existence of numerous archaeological and anthropological studies, which in one way or another have adopted ideas and techniques based on historical continuity – and inheritable continuity – and on the reconstruction of hypothetical chronologically ordered seriations. However, this first evolutionist phase shows important weaknesses. On the one hand, I must stress the fact that these approaches generally lacked an explicative theoretical position – or a theory explicitly linked to the phenomenon under study – on cultural change. For this reason, the majority of the authors cited above developed their classification schemes without formulating any specific questions regarding the nature of the cultural information that was being transferred over time and was materialized in the artifacts under study nor, regarding the ways, the mechanisms by which the information had been transmitted. On the other hand, a common characteristic of all of these seriations is that they were based on the criterion of the overall similarity between objects, for example, their form or the type of decoration that they displayed, rather than specifically on shared derived characters (synapomorphies) as is the case in modern cladistics.

In the 1960s, anthropological and archaeological studies shifted their focus away from the issues of cultural phylogenies and concentrated particularly on other aspects, often functionalist, from a processualism perspective. It was not until the 1980s that the interest in cultural phylogenies was renewed. This new phase thus owes part of its development to the processualist school, which invested much effort in the systematization and construction of models for the phenomena analyzed in the fields of anthropology and archaeology. The application of mathematics and statistics for the methodological purposes of modeling was greatly developed by the New Archaeology and this later became the underlying basis that facilitated the incorporation of phylogenetic techniques and methods into our field of knowledge. The advances in computer-based analytical methods of large volumes of data were, of course, also of great influence.

The weaknesses of the first evolutionist phase outlined above were partially remedied in the 1980's phase. From then onward, anthropology invested considerable efforts in the study of the processes involved in cultural transmission (e.g., Durham 1976, 1991; Pulliam and Dunford 1980; Cavalli-Sforza and Feldman 1981; Boyd and Richerson 1982, 1985; Richerson and Boyd 2005). It was also then that the theoretical debate on the general application of Darwinian theory in archaeology was reopened (e.g., Dunnell 1978, 1980; Rindos 1980, 1985; Borrero 1993; Teltser 1995; Maschner 1996; O'Brien 1996; Barton and Clark 1997; O'Brien and Lyman 2000; Hart and Terrell 2002; Shennan 2002). According to the available literature, the first (modern) cladistic application to archaeological materials may be traced back to that period, in the case of Iberian sculptures and La Tène fibulae, by T. Chapa (1980, 1984). Unfortunately, these studies were carried out in the Spanish academic sphere and remained isolated so that this line of enquiry was not followed up on at the time, not even by the author of those early studies.

The last developmental phase of phylogenetic methods in archaeology, leading to their current state, goes back to the past decade over which the main points of this methodology have been reviewed and defined in relation specifically to material culture (cf. O'Brien and Lyman 2000). As evidence of the present-day impetus of (now systematic) cultural phylogenetics, there are several recent publications that compile very varied technical, methodological, and thematic contributions to this field (Mace et al. 2005; Forster and Renfrew 2006; Lipo et al. 2006a; O'Brien 2008; Shennan 2009; Escacena et al. 2010).

As well as these monographic works, the past years have seen the proliferation of studies and applications of phylogenetic methods in publications of diverse characteristics and scope. Some of the most recognized specialists in this line of enquiry have suggested (cf. O'Brien et al. 2008: 40) that the phylogenetic studies carried out in anthropology and archaeology can be divided into three categories: (1) studies that track lines of transmission and descent back in time in search of common ancestors (prototypes) to examine the processes underlying the geographical distribution and cultural development of the descendants; (2) approaches that first create nested groups of related taxa, or clades, and then track those taxa geographically; and (3) comparative studies that depend on the understanding of models of descent in order to examine the distribution of functionally adaptive traits. In this line, following models developed in biology (e.g., Goodman et al. 1979; Page 1990; Page and Charleston 1998), some other cultural studies carry out co-phylogenetic approaches to investigate historical associations, that is, coevolutionary processes between different traits or taxa whose lineages are strongly associated with each other (e.g., Riede 2009). Furthermore, phylogenetic applications have been recently used to test specific interpretative historical hypotheses even of symbolic nature (e.g., García Rivero and O'Brien 2014).

These sorts of works, which systematically apply phylogenetic methods in research fields concerned with human cultural evolution, have dealt with very varied geographical settings and case studies from diverse chronological contexts in Europe, Asia, Africa, North America, and the Pacific (cf. Lipo et al. 2006b: 5).

In material culture studies, we find specific phylogenetic approaches as well as perspectives concerned primarily with the various processes of cultural transmission. Several references can be added to the list compiled by Lipo et al. (2006b: 5–6), including studies of lithic industries (Darwent and O'Brien 2006; Eerkens et al. 2006; Apel and Darmark 2007; Buchanan and Collard 2008), symbolic stone figurines (García Rivero 2010a; García Rivero and O'Brien 2014), bone industries (Riede 2008), pottery (Cochrane 2004, 2008; Harmon et al. 2006; Neff 2006), decorative elements on diverse materials (Vanpool et al. 2008), musical instruments (Tëmkin and Eldredge 2007), and combined traits in multidisciplinary approaches (Moylan et al. 2006; Coward et al. 2008).

## 6 Discussion

Applying the Darwinian theory of evolution in archaeology is not a common practice, although it has recently become more usual (Lycett 2015). Many scholars, particularly those who specialize in Pleistocene hominids, accept this theory only as long as it explains the biological evolution of anatomical features, but not when it is used to explain patterns of behavior or cultural materials, especially if these belong to *Homo sapiens* (Escacena 2010).

In contrast to North American scholarship, academic archaeology in Europe does not hold, in general, a strong tradition in the theoretical discussion of taxonomy. Despite the longstanding practice of classification, the employed approaches and methods remain biased and partial. Most European (and many American) archaeologists have put considerable effort into describing and recording every single trait considered in their studies. However, such effort has rarely gone beyond the descriptive task and very few researchers have used the recorded characters in analytical ways. In other words, the attributes that are supposed to constitute the core of typology have rarely been worked with operatively. This may be attributed to the lack of an underlying theoretical basis to enable appropriate methodological correlations. In addition, there is the exclusive existence of an essentialist phenetic typological conception of specimens without a theoretical consideration of other taxonomic concepts and methods.

The weak tradition in the study of taxonomic issues within European academia may be further explained by another methodological factor, influencing the work habits of researchers who have traditionally replicated the same approaches over generations. From its outset, North American archaeology was commonly faced with a decontextualized archaeological record, where external information on the stratigraphic position of finds or on the absolute chronology of the study objects was unavailable. In such a setting, discussions concerning archaeological taxonomies became fundamental. These revolved around the systematic analysis of traits and objects and the notion of historical, inheritable continuity. This approach continued successfully throughout the twentieth century. In the Old World, in contrast, the archaeological record was more substantial and contextual information for at least some of the best-known ancient sites was available from the beginning. The importance of stratigraphic laws became well-established in the mid-twentieth century[2] and was fully consolidated by the late 1970s. Also by then, absolute dating techniques had been refined, which constitutes a noteworthy factor. Reliance on stratigraphy and absolute dating led to the marginalization and near abandonment of techniques devoted to the intrinsic ordering of series of objects in European archaeology.

This observation does not intend to undermine the importance of stratigraphy and absolute dating techniques, which constitute basic tools in our field of study, but rather to highlight their renewed potential if they were to be combined with intrinsic taxonomic techniques based on historical continuity and kinship. The integration of all of these methods would then become a highly appropriate and fundamental resource for the formulation and the sequential analysis of hypotheses of historical nature.

The methodological and interpretative potential of phylogenetics in archaeology must be highlighted, being both fertile (leading to a considerable increase of historical hypotheses) and experimental (allowing comparison between hypotheses). In sum, the special relevance of cladistics may be synthesized in four points: (1) epistemic alignment with the Linnaean taxonomic model and the theory of evolution by descent with modification; (2) systematic application of objective and coherent principles, leading to the reconstruction of phylogenetic and kin relationships by means of nested groups of taxa based on shared derived characters; (3) possibility to test and to refute, i.e., the criterion of Popperian falsifiability; and (4) objective and systematic assessment of hypotheses and results by means of a statistic base that reduces the uncertainties and ambiguities of other archaeological methods and models.

Therefore, the potential and advantages of phylogenetic methodologies for archaeology should not be overlooked, and the (mistaken) criticisms of this line of enquiry can no longer be sustained. The only point that may constitute a founded objection to this method is the problem caused by horizontal transmission and homoplasy. As stated above, this is not a philosophical or theoretical issue but a strictly methodological question. In any case, an important argument against this criticism is that the cladistic method itself, as currently understood and used in its

[2] Even earlier within a broader academic framework, as illustrated by the phases established at Troy in the late nineteenth century.

most simple form, serves to assess possible models of transmission of information. That is, if our results indicate excessive phylogenetic noise, we can infer the signal of mechanisms by which some form of homoplasy is reflected by the data, and then we can question and consider the intensity of the different mechanisms of transmission that, in one way or another, affect the coherent cultural units under study (Jordan and Mace 2006: 151).

Future work may have to address the refinement of phylogenetic methodologies and the integration of the diverse techniques that have been developed over time, which I summarized in this contribution. As I noted above, these methods have undergone important changes in recent years, although mostly in other fields such as evolutionary biology (and particularly molecular biology), where they are used assiduously (e.g., Yang and Goldman 2008). Nevertheless, although the application of systematic phylogenetics in archaeology is relative recent, some of the most eminent specialists in this field have already noted that the phylogenetic methods of parsimony and maximum likelihood should be accompanied by other techniques, such as simulation, decomposition graphs and network analyses, serial independence tests, iterated parsimony, Bayesian methods like Monte Carlo chains, correspondence matrix analyses, and the assessment of the hierarchical structure of groups and seriation (O'Brien et al. 2008: 58). The effort will no doubt be rewarding.

**Acknowledgments** I would like to thank the University of Seville, the research group PAI-HUM949, and the project HAR2013-44269-P for their support and, also, the editors Nathalie Gontier, Olga Pombo, and Larissa Mendoza Straffon for their invitation to contribute to this book. Two anonymous reviewers made very interesting suggestions that contributed to improve this chapter.

## References

Aldenderfer, M. (2005). Statistics for archaeology. In H. D. G. Maschner, & C. Chippindale (Eds.), *Handbook of archaeological methods* (Vol. 1, pp. 501–553) Lanham: AltaMira Press.

Apel, J., & Darmark, K. (2007). Den flathuggna pilspetsens fylogeni. Mellansvenskt stenhantverk ur ett kulturevolutionistiskt perspektiv. In N. Stenbäck (Ed.), *Stenålder i Uppland. Uppdragsarkeologi och eftertanke* (pp. 31–65). Uppsala: Arkeologi E4.

Atkinson, Q. D., & Gray, R. D. (2005). Curious parallels and curious connections: Phylogenetic thinking in Biology and Historical Linguistics. *Systematic Biology, 54*(4), 513–526.

Atkinson, Q. D., & Gray, R. D. (2006). Are accurate dates an intractable problem for Historical Linguistics? In C. Lipo, et al. (Eds.) *Mapping our ancestors: Phylogenetic approaches in Anthropology and Prehistory* (pp. 269–296). New York: Aldine Transaction.

Baroni, M., Semple, C., & Steel, M. (2004). A framework for representing reticulate evolution. *Annals of Combinatorics, 8*, 391–408.

Baroni, M., Semple, C., & Steel, M. (2006). Hybrids in real time. *Systematic Biology, 55*(1), 46–56.

Barton, C. M., & Clark, G. A. (Eds). (1997). *Rediscovering Darwin: Evolutionary theory and archeological explanation*, Archeological papers of the American Anthropological Association 7. Arlington.

Bellwood, P. (1996). Phylogeny vs reticulation in prehistory. *Antiquity, 70*, 881–890.

Bentley, R. A., Lipo, C., Maschner, H. D. G., & Marler, B. (2008). Darwinian archaeologies. In A. Bentley, et al. (Eds.), *Handbook of archaeological theories* (pp. 109–132). Lanham: Altamira Press.

Borrero, L. A. (1993). Artefactos y evolución. *Palimpsesto. Revista de Arqueología, 3*, 15–32.
Boyd, R., & Richerson, J. R. (1982). Cultural transmission and the evolution of cooperative behaviour. *Human Ecology, 10*, 325–351.
Boyd, R., & Richerson, J. R. (1985). *Culture and the evolutionary process*. Chicago: Chicago University Press.
Breuil, H. (1935). *Les peintures rupestres schématiques de la Péninsule Ibérique*. Lagny-sur-Marne: Imprimerie de Laquy.
Bryant, D., & Moulton, V. (2002). NeighborNet: An agglomerative method for the construction of planar phylogenetic networks. In R. Guigó, & D. Gusfield (Eds.), *Workshop in algorithms for Bioinformatics*, Proceedings of the second international workshop in Rome – Lecture notes in computer science (pp. 375–391). Berlin & New York: Springer.
Bryant, D., Filimon, F., & Gray, R. D. (2005). Untangling our past: Languages, trees, splits and networks. In R. Mace, et al. (Eds.), *The evolution of cultural diversity. A phylogenetic approach* (pp. 67–84). London: College London University Press.
Buchanan, B., & Collard, M. (2008). Testing models of early paleoindian colonization and adaptation using Cladistics. In M. J. O'Brien (Ed.), *Cultural transmission and Archeology. Issues and case studies* (pp. 59–76). Washington, DC: Society for American Archaeology.
Carpenter, E., & Schuster, C. (1988). *Materials for the study of social symbolism in ancient and tribal art: A record of tradition and continuity*. New York: Foundation.
Cavalli-Sforza, L. L., & Feldman, M. W. (1981). *Cultural transmission and evolution: A quantitative approach*. Princeton, NJ: Princeton University Press.
Chapa, T. (1980). *La escultura zoomorfa ibérica en piedra* (2 Vols.), Doctoral Thesis, Universidad Complutense de Madrid.
Chapa, T. (1984). Aspectos metodológicos de la tipología arqueológica: Un ejemplo referido a las fíbulas de La Tene. In *Primeras jornadas de metodología de investigación prehistórica,* Soria, 1981 (pp. 253–268). Madrid: Ministerio de Cultura.
Clark, G. A., & Stafford, C. R. (1982). Quantification in American Archaeology: A Historical perspective. *World Archaeology, 14*(1), 98–119.
Cochrane, E. (2004). *Explaining cultural diversity in ancient Fiji: The transmission of ceramic variability*. Doctoral Thesis, Universidad de Hawaii.
Cochrane, E. (2008). Migration and cultural transmission: Investigating human movement as an explanation for Fijian ceramic change. In M. J. O'Brien (Ed), *Cultural transmission and Archeology. Issues and case studies* (pp. 132–145). Washington, DC: Society for American Archaeology.
Collard, M., & Tehrani, J. (2005). Phylogenesis versus Ethnogenesis in Turkmen cultural evolution. In R. Mace, et al. (Eds.), *The evolution of cultural diversity: A phylogenetic approach* (pp. 109–131). London: College London University Press.
Collard, M., Shennan, S. J., & Tehrani, J. (2006). Branching versus blending in macroscale cultural evolution: A comparative study. In C. Lipo, et al. (Eds.), *Mapping our ancestors: Phylogenetic approaches in Anthropology and Prehistory* (pp. 53–64). New York: Aldine Transaction.
Colton, H. S., & Hargrave, L. L. (1937). Handbook of northern Arizona pottery wares. Museum of Northern Arizona, 11. Cowgill, G. L.
Coward, F., Shennan, S. J., Colledge, S., Conolly, J., & Collard, M. (2008). The spread of Neolithic plant economies from the Near East to northwest Europe: A phylogenetic analysis. *Journal of Archaeological Science, 35*, 42–56.
Currie, T. E., Greenhill, S. J., & Mace, R. (2010). Is horizontal transmission really a problem for phylogenetic comparative methods? A simulation study using continuous cultural traits. *Philosophical Transactions of the Royal Society B, 365*, 3903–3912.
Darwent, J., & O'Brien, M. J. (2006). Using Cladistics to construct lineages of projectile points from Northeastern Missouri. In C. Lipo, et al. (Eds), *Mapping our ancestors: Phylogenetic approaches in Anthropology and Prehistory* (pp. 185–208). New York: Aldine Transaction.
Darwin, C. R. (1859). *On the origin of species by means of natural selection*. London: John Murray.
Diamond, J. (1999). *Guns, germs, and steel. The fates of human societies*. New York: W.W. Norton & Company.

Doran, J. E., & Hodson, F. R. (1975). *Mathematics and computers in Archaeology*. Edinburgh: Edinburgh University Press.
Dunnell, R. C. (1978). Style and function: A fundamental dichotomy. *American Antiquity, 43*(2), 192–202.
Dunnell, R. C. (1980). Evolutionary theory and archaeology. *Advances in Archaeological Method and Theory, 3*, 35–99.
Dunnell, R. C. (1982). Science, social science, and common sense: The agonizing dilemma of modern archaeology. *Journal of Archaeological Research, 38*, 1–25.
Durham, W. H. (1976). The adoptive significance of cultural behaviour. *Human Ecology, 4*, 89–117.
Durham, W. H. (1991). *Coevolution: Genes, culture and human diversity*. Stanford, CA: Stanford University Press.
Eerkens, J. W., Bettinger, R. L., & McElreath, R. (2006). Cultural transmission, Phylogenetics, and the archaeological record. In C. Lipo, et al. (Eds.), *Mapping our ancestors: Phylogenetic approaches in Anthropology and Prehistory* (pp. 169–183). New York: Aldine Transaction.
Escacena, J. L. (2010). Darwinismo en almíbar. In Escacena, et al. (Eds.), *Clasificación y Arqueología: Enfoques y métodos taxonómicos a la luz de la evolución darwiniana* (pp. 33–57). Sevilla: Universidad de Sevilla.
Escacena, J. L., García Rivero, D., & García Fernández, F. J. (Eds.) (2010). *Clasificación y Arqueología: Enfoques y métodos taxonómicos a la luz de la evolución darwiniana*. Sevilla: Universidad de Sevilla.
Farris, J. S. (1970). Methods for computing Wagner trees. *Systematic Zoology, 19*, 83–92.
Fernández Martínez, V. M. (1985). La seriación automática en Arqueología: Introducción histórica y aplicaciones. *Trabajos de Prehistoria, 42*, 9–49.
Forster, P., & Renfrew, C. (Eds). (2006). *Phylogenetic methods and the prehistory of languages*. Cambridge: McDonald Institute Monographs.
García Rivero, D. (2010a). *Evolución cultural y filogenias en Arqueología. El caso de los denominados ídolos placa prehistóricos del Suroeste de la Península Ibérica*. Doctoral Thesis, Universidad de Sevilla.
García Rivero, D. (2010b). Introducción a la teoría de la clasificación y a las escuelas taxonómicas (Fenética, Cladística y Taxonomía evolutiva). In Escacena, et al. (Eds.), *Clasificación y Arqueología: Enfoques y métodos taxonómicos a la luz de la evolución darwiniana* (pp. 61–94). Sevilla: Universidad de Sevilla.
García Rivero, D. (2013). *Arqueología y evolución. A la búsqueda de filogenias culturales*. Sevilla: Universidad de Sevilla.
García Rivero, D., & O'Brien, M. J. (2014). Phylogenetic analysis shows that Neolithic slate plaques from the Southwestern Iberian Peninsula are not genealogical recording systems. *PLoS ONE,* 9(2), e88296.
Goloboff, P. A. (2003). Parsimony, likelihood, and simplicity. *Cladistics, 19*, 91–103.
Goodman, M. J., Czelusniak, J., Moore, G. W., Romero-Herrera, A. E., & Matsuda, G. (1979). Fitting the gene lineage into its species lineage: A parsimony strategy illustrated by cladograms constructed from globin sequences. *Systematic Zoology, 28*, 132–168.
Gould, S. J. (1987). *An urchin in the storm*. New York: WW Norton & Company.
Gray, R. D., & Atkinson, Q. D. (2003). Language tree divergence times support the Anatolian theory of Indo-European origin. *Nature, 426*, 435–439.
Gray, R. D., & Jordan, F. (2000). Language trees support the Express-train sequence of Austronesian expansion. *Nature, 405*, 1052–1055.
Gray, R. D., Greenhill, S. J., & Ross, R. M. (2007). The pleasures and perils of darwinizing culture (with phylogenies). *Biological Theory, 2*(4), 360–375.
Greenhill, S. J., & Gray, R. D. (2005) Testing population dispersal hypotheses: Pacific settlement, phylogenetic trees and Austronesian languages. In R. Mace, et al. (Eds), *The evolution of cultural diversity: A phylogenetic approach* (pp. 31–52). London: College London University Press.

Greenhill, S. J., Currie, T. E., & Gray, R. D. (2009). Does horizontal transmission invalidate cultural phylogenies? *Proceedings of the Royal Society B, 276*, 2299–2306.

Haig, D. (2007). Weismann Rules! Ok? Epigenetics and the Lamarckian temptation. *Biology and Philosophy, 22*, 415–428.

Harmon, M. J., Vanpool, T. L., Leonard, R. D., Vanpool, C. S., & Salter, L. A. (2006). Reconstructing the flow of information across time and space: A phylogenetic analysis of ceramic traditions from prehispanic Western and Northern Mexico and the American Southwest. In C. Lipo, et al. (Eds.), *Mapping our ancestors: Phylogenetic approaches in Anthropology and Prehistory* (pp. 209–229). New York: Aldine Transaction.

Hart, J. P., & Terrell, J. E. (Eds.). (2002). *Darwin and Archaeology. A handbook of key concepts*. Westport, CT: Bergin & Garvey.

Hennig, W. (1950). *Grundzüge einer Theorie der phylogenetischen Systematik*. Berlin: Deutscher Zentralverlag.

Hennig, W. (1965). Phylogenetic systematics. *Annual Review of Entomology, 10*, 97–116.

Hennig, W. (1966). *Phylogenetic systematics*. Illinois: Illinois University Press.

Holden, C. J. (2006). The spread of Bantu languages, farming, and pastoralism in Sub-Equatorial Africa. In C. Lipo, et al. (Eds.), *Mapping our ancestors: Phylogenetic approaches in Anthropology and Prehistory* (pp. 249–267). New York: Aldine Transaction.

Holden, C. J., Meade, A., & Pagel, M. (2005). Comparison of maximum parsimony and Bayseian Bantu language trees. In R. Mace, et al. (Eds), *The evolution of cultural diversity: A phylogenetic approach* (pp. 53–65). London: College London University Press.

Holder, M., & Lewis, P. O. (2003). Phylogeny estimation: Traditional and Bayesian approaches. *Nature Reviews Genetics, 4*(4), 275–284.

Jones, W. (1786). Third anniversary discourse 'On the Hindus'. Reedited in *The Collected Works of Sir William Jones*, 1807 (pp. 23–46). London: John Stockdale.

Jordan, P., & Mace, T. (2006). Tracking culture-historical lineages: Can 'descendent with modification' be linked to 'association by descent'? In C. Lipo, et al. (Eds.), *Mapping our ancestors: Phylogenetic approaches in Anthropology and Prehistory* (pp. 149–167). New York: Aldine Transaction.

Jordan, P., & O'Neill, S. (2010). Untangling cultural inheritance: Language diversity and longhouse architecture on the Pacific Northwest Coast. *Philosophical Transactions of the Royal Society B, 365*, 3875–3888.

Kitching, I. J., Forey, P. L., Humphries, C. J., & Williams, D. J. (1998) *Cladistics. The theory and practice of parsimony analysis*. Oxford: Systematics Association Publication 11.

Lamarck, J. B. (1809). *Philosophie zoologique*. Paris: Dentu.

Laskowski, M., & Fitch, W. M. (1989). Evolution of avian ovomucoids and of birds. In B. Fernholm, et al. (Eds.), *The Hierarchy of life* (pp. 371–387). New York & Oxford: Elsevier Science Publishers.

Lewis, P. O. (2001). Phylogenetic systematics turns over a new leaf. *Trends in Ecology and Evolution, 16*(1), 30–37.

Lillios, K. (2004). Esprit (The engraved stone plaque registry and inquiry tool). http://research2.its.uiowa.edu/iberian/. Accessed 15 Nov 2013.

Lipo, C. P., O'Brien, M. J., Collard, M., & Shennan, S. J. (Eds.). (2006a). *Mapping our ancestors: Phylogenetic approaches in Anthropology and Prehistory*. New York: Aldine Transaction.

Lipo, C. P., O'Brien, M. J., Collard, M., & Shennan, S. J. (2006b). Cultural phylogenies and explanation: Why historical methods matter. In C. Lipo, et al. (Eds.), *Mapping our ancestors: Phylogenetic approaches in Anthropology and Prehistory* (pp. 3–16). New York: Aldine Transaction.

Lycett, S. J. (2015). Cultural evolutionary approaches to artifact variation over time and space: Basis, progress, and prospects. *Journal of Archaeological Science, 56*, 21–31.

Lyman, R. L., & O'Brien, M. J. (2006). Seriation and Cladistics: The difference between anagenetic and cladogenetic evolution. In C. Lipo, et al. (Eds.), *Mapping our ancestors: Phylogenetic approaches in Anthropology and Prehistory* (pp. 65–88). New York: Aldine Transaction.

Mace, R., Holden, C. J., & Shennan, S. J. (Eds.). (2005). *The evolution of cultural diversity: A phylogenetic approach*. London: College London University Press.
Madsen, T. (1988). *Multivariate archaeology: Numerical approaches in Scandinavian archaeology*. Højbjerg: Jutland Archaeological Society.
Makarenkov, V., & Legendre, P. (2004). From a phylogenetic tree to a reticulated network. *Journal of Computational Biology, 11*(1), 195–212.
Margulis, L., & Sagan, D. (2002). *Acquiring genomes: A theory of the origins of species*. New York: Basic Books.
Maschner, H. D. G. (Ed). (1996). *Darwinian archaeologies*. New York: Plenum Press.
Mayr, E. (1969). *Principles of systematic Zoology*. New York: McGraw-Hill.
Mayr, E., & Ashlock, P. D. (1991). *Principles of systematic Zoology* (2nd ed.). Singapore: McGraw Hill International Edition.
McDade, L. A. (1992). Hybrids and phylogenetic systematics II. The impact of hybrids on cladistic analysis. *Evolution, 46*(5), 1329–1346.
McGlade, J. (2005). Systems and simulacra: Modeling, simulation, and archaeological interpretation. In H. D. G. Maschner, & C. Chippindale (Eds.), *Handbook of archaeological methods* (Vol. 1, pp. 554–602). Lanham: AltaMira Press.
McMullin, E. (1983). Values in science. *Proceedings of the Biennial Meeting of the Philosophy of Science Association,* 1982(2), 3–28.
Mesoudi, A., Whiten, A., & Laland, K. N. (2006). Towards a unified science of cultural evolution. *Behavioral and Brain Sciences, 29*(4), 329–347.
Moore, J. H. (1994). Putting Anthropology back together again: The ethnogenetic critique of cladistic theory. *American Anthropologist, 96*, 925–948.
Morgan, L. H. (1877). *Ancient society*. New York: Holt.
Moylan, J. W., Graham, C. M., Borgerhoff, M., Nunn, C. L., & Hakansson, N. T. (2006). Cultural traits and linguistic trees: Phylogenetic signal in East Africa. In C. Lipo, et al (Eds.) *Mapping our ancestors: Phylogenetic approaches in Anthropology and Prehistory* (pp. 33–52). New York: Aldine Transaction.
Muscio, H. J. (2010). Transferencia horizontal, cladismo y filogenias culturales. In Escacena, et al. (Eds.), *Clasificación y Arqueología: Enfoques y métodos taxonómicos a la luz de la evolución darwiniana* (pp. 225–252). Sevilla: Universidad de Sevilla.
Neff, H. (2006). Archaeological-material characterization as phylogenetic method: The case of Copador pottery from Southeastern Mesoamerica. In C. Lipo, et al. (Eds.), *Mapping our ancestors: Phylogenetic approaches in Anthropology and Prehistory* (pp. 231–248). New York: Aldine Transaction.
Nicholls, G., & Gray, R. (2008). Dated ancestral trees from binary trait data and their application to the diversification of languages. *Journal of the Royal Society B, Statistical Methodology, 70*(3), 545–566.
O'Brien, M. J. (Ed.). (1996). *Evolutionary archaeology. Theory and application*. Salt Lake City: Utah University Press.
O'Brien, M. J. (Ed.). (2008). *Cultural transmission and Archaeology. Issues and case studies*. Washington, DC: Society for American Archaeology.
O'Brien, M. J., & Lyman, R. L. (1999). *Seriation, stratigraphy, and index fossils: The backbone of archaeological dating*. New York: Kluwer Academic y Plenum.
O'Brien, M. J., & Lyman, R. L. (2000). *Applying evolutionary archaeology. A systematic approach*. New York: Kluwer Academic y Plenum Publishers.
O'Brien, M. J., & Lyman, R. L. (2003). *Cladistics and archaeology.* Salt Lake City: Utah University Press.
O'Brien, M. J., Holland, T. D., & Fox, G. L. (2003). Evolutionary implications of design and perfomance characteristics of prehistoric pottery. In M. J. O'Brien, & R. L. Lyman (Eds.), *Syle, fuction, transmission. Evolutionary archaeological perspectives*, Foundations of Archaeological Inquiry (pp. 199–226). Salt Lake City: Utah University Press.

O'Brien, M. J., Lyman, R. L., Collard, M., Holden, C. J., Gray, R. D., & Shennan, S. (2008). Transmission, Phylogenetics, and the evolution of cultural diversity. In M. J. O'Brien (Ed.), *Cultural transmission and archeology. Issues and case studies* (pp. 39–58). Washington, DC: Society for American Archaeology.

O'Hara, R. J. (1997). Population thinking and tree thinking in systematics. *Zoologica Scripta, 26*(4), 323–329.

Page, R. D. (1990). Temporal congruence and cladistic analysis of biogeography and cospeciation. *Systematic Zoology, 39*, 205–226.

Page, R. D., & Charleston, M. A. (1998). Trees within trees: Phylogeny and historical associations. *Trends in Ecology and Evolution, 13*, 356–359.

Platnick, N. I., & Cameron, H. D. (1977). Cladistic methods in textual, linguistic, and phylogenetic analysis. *Systematic Biology, 26*(4), 380–385.

Pulliam, H. R., & Dunford, C. (1980). *Programmed to learn: An essay on the evolution of culture*. New York: Columbia University Press.

Renfrew, C. (1992). Archaeology, genetics and linguistic diversity. *Man, 27*, 445–478.

Rexová, K., Frynta, D., & Zrzavy, J. (2003). Cladistic analysis of languages: Indo-European classification based on lexicostatistical data. *Cladistics, 19*, 120–127.

Richerson, P. J., & Boyd, R. (2005) *Not by genes alone. How culture transformed human evolution*. Chicago y London: Chicago University Press.

Riede, F. (2008). Maglemosian memes: Technological ontogeny, craft traditions, and the evolution of Northern European barbed points. In M. J. O'Brien (Ed.), *Cultural transmission and archeology. Issues and case studies* (pp. 178–189). Washington, DC: Society for American Archaeology.

Riede, F. (2009). Tangled trees: Modelling material culture evolution as host-associate cospeciation. In S. J. Shennan (Ed.), *Pattern and process in cultural evolution* (pp. 85–98). Berkeley: University of California Press.

Rindos, D. (1980). Symbiosis, instability, and the origins and spread of agriculture: A new model. *Current Anthropology, 21*(6), 751–772.

Rindos, D. (1985). Darwinian selection, symbolic variation and the evolution of culture. *Current Anthropology, 26*(1), 65–88.

Rivera, M. C., & Lake, J. A. (2004). The ring of life provides evidence for a genome fusion origin of eukaryotes. *Nature, 431*, 152–155.

Ronquist, F. (2004). Bayesian inference of character evolution. *Trends in Ecology and Evolution, 19*(9), 475–481.

Ruse, M. (1999). *Mystery of mysteries. Is evolution a social construction?* Cambridge, MA: Harvard University Press.

Sanderson, M. J., & Donoghue, M. J. (1989). Patterns of variation in levels of homoplasy. *Evolution, 43*(8), 1781–1795.

Schiffer, M. B. (2008). Transmission processes: A behavioral perspective. In M. J. O'Brien (Ed.), *Cultural transmission and archeology. Issues and case studies* (pp. 102–111). Washington, DC: Society for American Archaeology.

Schleicher, A. (1863). *Die Darwinsche Theorie und die Sprachwissenschaft*. Weimar: Böhlau.

Service, E. R. (1962). *Primitive social organization*. New York: Random House.

Shennan, S. J. (2002). *Genes, memes and human history. Darwinian archaeology and cultural evolution*. London: Thames & Hudson.

Shennan, S. J. (Ed.). (2009). *Pattern and process in cultural evolution*. Berkeley: University Press.

Spencer, M., Windram, H. F., Barbrook, A. C., Davidson, E. A., & Howe, C. J. (2006). Phylogenetic analysis of written traditions. In P. Forster, & C. Renfrew (Eds.) *Phylogenetic methods and the prehistory of languages* (pp. 67–74). Cambridge: McDonald Institute Monographs.

Steward, J. (1955). *Theory of culture change*. Urbana: Illinois University Press.

Tehrani, J. J., & Riede, F. (2008). Towards an archaeology of pedagogy: Learning, teaching and the generation of material culture traditions. *World Archaeology, 40*(3), 316–331.

Teltser, P. A. (Ed.). (1995). *Evolutionary archaeology: Methological issues*. Tucson: Arizona University Press.

Tëmkin, I., & Eldredge, N. (2007). Phylogenetics and material cultural evolution. *Current Anthropology, 48*(1), 146–153.
Terrell, J. E., & Steward, P. J. (1996). The paradox of human population genetics at the end of the twentieth century. *Reviews in Anthropology, 26*, 13–33.
Tylor, E. B. (1871). *Primitive culture*. London: Murray.
Van Driem, G. (2001). *Languages of the Himalayas: An ethnolinguistic handbook of the Greater Himalayan region* (2 Vols.). Leiden: Brill.
Vanpool, T. L., Palmer, C. T., & Vanpool, C. S. (2008). Horned serpents, tradition, and the tapestry of culture. In M. J. O'Brien (Ed.), *Cultural transmission and archeology. Issues and case studies* (pp. 77–90). Washington, DC: Society for American Archaeology.
White, L. A. (1959). *The evolution of culture*. New York: McGraw-Hill.
Willey, E. O., Siegel-Caysey, D., Brooks, D. R., & Funk, V. A. (1991). *The compleat cladist. A primer of phylogenetic procedures*. The University of Kansas, Museum of Natural History, Special Publication 19. Kansas: Lawrence.
Woese, C. R. (2004). A new biology for a new century. *Microbiology and Molecular Biology Reviews, 68*(2), 173–186.
Yang, Z, & Goldman, N. (2008). *Statistical and computational challenges in molecular Phylogenetics and evolution*. London: Royal Society Publishing.

# The Importance of a "Quantitative Genetic" Approach to the Evolution of Artifact Morphological Traits

**Stephen J. Lycett**

**Abstract** Recent years have seen substantial growth in the application of evolutionary approaches to spatial and temporal variation exhibited in archaeological data. As is now well known, the application of this approach rests on the basis that artifacts are an expression of a genuine evolutionary system mediated by transmission (via social learning), variation in transmitted elements, and differential replication of transmitted elements across time. While this provides the necessary fundamental basis for the application of an evolutionary approach to artifactual variation, application of the term "evolution" still provides a source of confusion for some archaeologists. Part of this confusion may stem from an underdeveloped body of theory that conceptually makes explicit the link between the evolution of socially transmitted information and the *expression* of that evolutionary process in terms of physical artifacts. This is especially the case given that artifactual variation is inevitably influenced by several different factors (e.g., raw material properties and/or post manufacture attrition), not all of which are necessarily heritable in systems of social learning. In order to resolve these difficulties and make more clear the case for an evolutionary approach to artifactual variation, there is a need for an explicit quantitative body of theory that links statistical variation in artifactual traits to factors such as selection and drift when (1) sources of artifact variation are multiple and not all necessarily heritable, (2) the proximate socially transmitted elements are unknown, and (3) many artifactual traits will be influenced simultaneously by multiple aspects of socially transmitted practices. Here, it is argued that a "quantitative genetic" approach can resolve these problems.

**Keywords** Cultural evolution • Evolutionary archaeology • Artifact evolution • Heritability • Social learning

---

S.J. Lycett
Department of Anthropology, University at Buffalo, SUNY
380 MFAC-Ellicott Complex, Buffalo, NY 14261, USA
e-mail: sjlycett@buffalo.edu



L. Mendoza Straffon (ed.), *Cultural Phylogenetics: Concepts and Applications in Archaeology*, Interdisciplinary Evolution Research,
DOI 10.1007/978-3-319-25928-4_4

## 1 Introduction

As others have noted (e.g., O'Brien and Lyman 2000; Shennan 2006), the roots of applying evolutionary principles to archaeological data go back to the late 1800s and early 1900s, during what some would now refer to as archaeology's "cultural historical" phase of intellectual development (Trigger 1989). Several have also shown that evolutionary thinking was an integral part of David Clarke's (1968) *Analytical Archaeology*, even if he was not always explicit on this point (e.g., O'Brien and Lyman 2000; Shennan 2004; Lycett and Chauhan 2010a; Lycett 2013). The potential for more explicit applications of evolutionary principles to archaeological questions began to be explored more earnestly, however, from the late 1970s onward (e.g., Dunnell 1978, 1980; Leonard and Jones 1987; Rindos 1989; O'Brien and Wilson 1988; O'Brien and Holland 1990). Also at this time, seminal works by authors outside of archaeology would begin to acknowledge the role that archaeological data might play in understanding human cultural phenomena in evolutionary terms (Cavalli-Sforza and Feldman 1981), and key works on social transmission were published (e.g., Boyd and Richerson 1985) upon which later applications of evolutionary archaeology would draw heavily. However, between the late 1970s and early 1990s, a series of what have been rightly referred to as "benchmark" papers were published (O'Brien 1996a: xiii), which explicitly began the brave task of trying to convince an archaeological audience that principles originally developed to study biological evolution might also be applicable to the archaeological record.

It is now almost two decades since many of these pioneer papers were conveniently brought together under a single published cover (O'Brien 1996b). However, it would be difficult for any archaeologist to be unaware of the considerable recent expansion of archaeological studies now explicitly utilizing evolutionary principles and methods. Importantly, one feature of this new generation of studies is their highly empirical nature, something which some of its earliest proponents have pointed out was sorely needed if the field was to progress (O'Brien and Lyman 2000: 22). For example, O'Brien and colleagues' (2001; O'Brien and Lyman 2003) application of quantitative phylogenetic methods to archaeological data has now been followed by a wide range of formal phylogenetic studies looking at issues of technological relatedness, diversification, and convergence in artifactual data (e.g., Tehrani and Collard 2002; Jordan and Shennan 2003, 2009; Lycett 2007, 2009a; Buchanan and Collard 2008; Rogers et al. 2009; Cochrane and Lipo 2010; Jordan and O'Neill 2010; Matthews et al. 2011; Marwick 2012; Jennings and Waters 2014) as well as applications of these methods to specifically examine questions of human dispersal (Buchanan and Collard 2007; Lycett 2009b). Moreover, archaeological data has been employed to track historical factors associated with the social transmission and the differential persistence of artifactual variation across time and space, again drawing explicitly on evolutionary theory and/or techniques of analysis to achieve this aim (e.g., Lipo et al. 1997; Bettinger and Eerkens 1999; Shennan and Wilkinson 2001; Vaughan 2001; Van Pool 2001; Bentley and Shennan 2003; Bentley et al. 2004; Kohler et al. 2004; Eerkens and Lipo 2005; Lycett and von

Cramon-Taubadel 2008, 2015; Lycett 2008; Lyman et al. 2008; Rogers and Ehrlich 2008; Mesoudi and O’Brien 2008; Buchanan and Hamilton 2009; Hamilton and Buchanan 2009; Brantingham and Perreault 2010; Lipo et al. 2010; Steele et al. 2010; Kempe et al. 2012; Cochrane et al. 2013; Okumura and Araujo 2014). It is perhaps important to note that these evolutionary studies of material culture have covered time periods from the Paleolithic through to ethnographically and historically recorded items and have dealt with artifact classes from across the globe as diverse as pottery, stone tools, baskets, carpets, house architecture, and watercraft. These above-listed studies, which have largely been published in scientific journals, have been joined by a series of edited volumes that either in whole or in part also deal with the application of evolutionary theory and methods to archaeological data (e.g., Hurt and Rakita 2001; Mace et al. 2005; Lipo et al. 2006; O’Brien 2008; Shennan 2009; Lycett and Chauhan 2010b; O’Brien and Shennan 2010; Ellen et al. 2013). The present volume, of course, adds to this growing list.

Collectively, in building on earlier efforts, this body of more recent research has illustrated two key points. Firstly, the diversity of important anthropological questions to which evolutionary studies of artifacts can, and indeed must be, applied. Secondly, in so doing, they have illuminated the powerful potential of the archaeological record to shed direct light on issues of social inheritance, the existence of variation, and the differential replication and persistence of those variants over the dimensions of time and space. That is, they have reemphasized that the archaeological record is both a pattern and problem of dealing with “descent with modification,” which makes it inherently an *evolutionary* issue (O’Brien and Lyman 2000; Lycett 2011; Mesoudi 2011).

Such a burgeoning literature might be taken to indicate a profound recent “success” in the acceptance of evolutionary approaches within archaeology, and to some extent, this is certainly correct. However, despite such apparent attainment of legitimacy, it is still not unusual to encounter strong viewpoints arguing that such approaches are, in effect, baseless. Indeed, personal experience at major international conferences has led to recent situations where even highly accomplished colleagues within the profession are not necessarily coy about making statements such as “phylogenetics cannot be applied to artifacts because artifacts don’t have genes” or, similarly, “artifacts cannot evolve because they are inanimate objects.” Such criticisms will be familiar to anyone seriously engaged in the evolutionary analysis of artifactual variation, who have repeatedly pointed out the fallacious (and jaded) character of such arguments, which can actually be traced back quite some time to at least Brew’s (1946) assertion that “pots don’t breed” (e.g., see O’Brien and Lyman 2000: 9).

One particular reason why evolutionary approaches to artifactual variation may cause confusion is the practical disconnect between the evolutionary process and the *expression* of that evolutionary process in the form of artifacts. Anyone tempted to think, for example, that “evolution is inapplicable to artifacts because they are inanimate objects” might well be reminded that from the standpoint of biological evolution, the physical body in which they reside is an inanimate object; their body will not itself evolve, but that does not mean it is not the expression of an evolutionary process mediated by factors of drift or selection, nor that it cannot play a role in the future of the evolutionary process (e.g., be a focus of selection). In strict terms, it is genetic *information systems* that evolve in the case of biological evolution. Hence, put bluntly, the skull of an individual horse can no more evolve than a pot used to carry water, but that does not repudiate that both are expressions of evolving *information systems* or, most importantly, that tracking their variation over time and space cannot reveal important information about that underlying evolutionary process (see, e.g., O'Brien and Lyman 2000; Eerkens and Lipo 2007; O'Brien et al. 2010). In the case of biology, that information system is coded at the molecular level in "genes," while the *cultural information system* is comprised of socially transmitted ideas, concepts, beliefs, and/or practices that either consciously or otherwise influence the form of the archaeological record at a particular time and geographic locality. A variety of different social transmission pathways of varying form and complexity may be involved in this process (for reviews of different social learning mechanisms, see, e.g., Byrne and Russon 1998; Whiten et al. 2004). For example, information about some element of an activity involving tools may be drawn to the attention of another because of usage, causing that latter individual to also adopt that behavior (so-called *stimulus enhancement*). Alternatively, it might be information about the manner in which a particular artifact looks and that form is later copied (*emulation*), and/or it might involve repeating actual behavioral details (e.g., hand position during pottery production) which are then, in turn, copied (*imitation*). It may also, of course, involve an individual directly and deliberately guiding the attention and/or behavior of another such that the particular artifact is replicated more readily (what many would term *teaching*) (e.g., Thornton and Raihani 2010). Combinations of these mechanisms are also feasible.

Perhaps the history of the development of biological evolution within scientific endeavor should tell us, however, that an understanding of the basic (albeit fundamental) components of a legitimate evolutionary approach is not alone necessarily enough to convince all of its necessity. As Bowler (2003:325) has detailed "[s]cientists had received Darwin's theory of natural selection with many reservations, and in the early years of the twentieth century, the level of hostility increased." Proponents of an evolutionary approach to artifactual variation have repeatedly pointed out that tackling questions concerning the historical, temporal, and spatial dynamics of the archaeological record is entirely legitimate because it has all three key components of a genuine system of "descent with modification" (e.g., O'Brien and Lyman 2000; Shennan 2000; Mesoudi et al. 2004; Lycett 2010). However, if Darwin's (1859) incredible insights concerning the long-term effects of variation, inheritance, and the biased replication of subsets of those variants were not enough to quell forcible pockets of resistance to key elements of his theory, perhaps we might be less surprised that merely (re)stating the existence of these three essential components in cultural systems will also be insufficient to convince everyone of the equally legitimate nature of this approach to artifactual data. As accepted as Darwin's (1859) outlining of the process of "descent with modification" now is within biology, much hard work was done in both theoretical and practical terms over the intervening decades to explain the *statistical mechanics* of evolution as they were applied to physical traits (e.g., Fisher 1918; Wright 1931; Mather 1943; Falconer 1960). Descriptions of these evolutionary mechanics became especially important in order to more precisely connect statistical variations in the physical traits of plants and animals to evolutionary processes such as selection and drift given increased recognition of the complexity of factors affecting their variation. In particular, three key points were increasingly recognized: (1) that traits of organisms are influenced by both heritable and nonheritable factors (e.g., "environmental" factors such as nutrition), (2) that individual quantitative traits of organisms (e.g., "height") were simultaneously influenced by several different heritable elements, and (3) that while traits were influenced by particulate inheritance according to Mendelian principles, patterns of variation displayed by the majority of phenotypic characters were continuously distributed in form (Provine 1971; Roff 1997).

As many have pointed out (e.g., O'Brien and Lyman 2000; Mesoudi et al. 2004; Lycett 2011), Darwin's (1859), characterization of evolution as a process of "descent with modification" does not depend on how traits are passed on, which is why a theoretical framework originally designed to help explore biological phenomena can be applied to cultural phenomena with equal legitimacy. However, given the circuitous relationship between the evolution (i.e., descent with modification) of the underlying information system and its physical expression in the form of artifacts, a fully developed evolutionary approach requires a precise (i.e., quantitative) framework that outlines—in specific terms—how patterns observed in physical forms over time and space respond to evolutionary forces such as drift and selection when individual objects are not themselves the entities that are evolving. This is especially the case when the exact form of these physical objects is influenced not just by heritable information (either socially learned concepts or genetically transmitted coding for phenotypes) but also might be influenced by factors outside the underlying information systems (e.g., "environment"). Only with the precise outlining and reconciliation of these factors did the full weight of Darwin's foundational ideas reach their now exalted status within biology (Provine 1971). This was especially so in terms of how selective and stochastic forces (i.e., drift) are linked mechanistically to subtle variations in the different traits of physical forms that are, in effect, by-products (or one might even say "artifacts") of the evolutionary process they document (Mayr 1982; Bowler 2003). In biology, these matters were resolved by the development of a field that eventually became known as "quantitative genetics" (e.g., Falconer and Mackay 1996; Roff 1997). This approach was necessary in order to translate insights provided by theoretical population genetics (i.e., the statistical study of genetic evolution, sometimes in hypothetical terms) more directly into a practical framework for studying the more unruly *physical* traits of organisms actually measurable in wild populations, especially when the exact details of the underlying genetic coding of these traits are (typically) completely unknown (Conner and Hartl 2004: 30; Roff 2007).

Recently, a similar framework has been proposed and begun to be developed as a means of tackling analogous problems inherent to the evolutionary analysis of material artifacts (Lycett and von Cramon-Taubadel 2015). The following section of this chapter will explain the fundamental elements of this approach. Thereafter, these factors will be expanded upon, in quantitative terms, to show more precisely how patterns of variation observed in the physical attributes of artifacts over time and space can be linked directly to evolutionary forces such as selection and stochastic factors when (1) sources of artifact variation are multiple and not all necessarily heritable, (2) the proximate socially transmitted elements are unknown, and (3) many artifactual traits will be influenced simultaneously by multiple aspects of socially transmitted practices. This approach is necessary, just as it was in biology, in order to provide more clear mechanistic links between the process of evolution and the statistical properties of its measurable physical record.

## 2 Quantitative Genetics: The Basis for an Evolutionary Approach to (Unruly) Continuous Physical Traits

A modern definition of biological evolution might define it as "a change in the gene pool" (e.g., Dawkins 1989: 45); a definition which emphasizes that biological "descent with modification" has a proximate molecular basis. However, phenotypic traits such as skull length, the shape of a beak, or the diameter of a tooth are not genes. This is despite the fact that when it comes to understanding the evolution of those features in physical terms, and determining the evolutionary processes at work, morphometric variation obviously comprises a key component of the basic data. As the Mendelian (particulate) basis of inheritance of such traits began to be properly recognized during the early part of the twentieth century, these matters caused a series of major problems for evolutionary biology (see, e.g., Provine 1971; May 1982; Bowler 2003 for historical overviews). This was compounded as experiments began to demonstrate for the first time that variation between individuals was caused not just by heritable components but also by "environmental" factors, such as the soil conditions within which a plant grew (e.g., Johannsen 1909). If the statistical study of physical traits was to form a notable component of the study of evolution, then these problems needed to be resolved. Moreover, they needed to be solved in a framework that did not rely on knowing, in absolute terms, which genes were responsible for which traits; in other words, it needed a "black box" approach to the study of evolution via statistical study of variation in physical traits (Conner and Hartl 2004: 103).

What is now known as the field of "statistical quantitative genetics" provided the urgently needed resolution to these problems (Roff 1997, 2007). A key element in the development of quantitative genetics was Ronald Fisher's (1918) insight that unlike variations between different types of simple differences in classic "Mendelian" traits (such as differences in flower color or between round versus wrinkled forms of pea), which were caused by allelic differences at a single genetic locus, variation in quantitative traits was caused by segregation of their heritable properties across

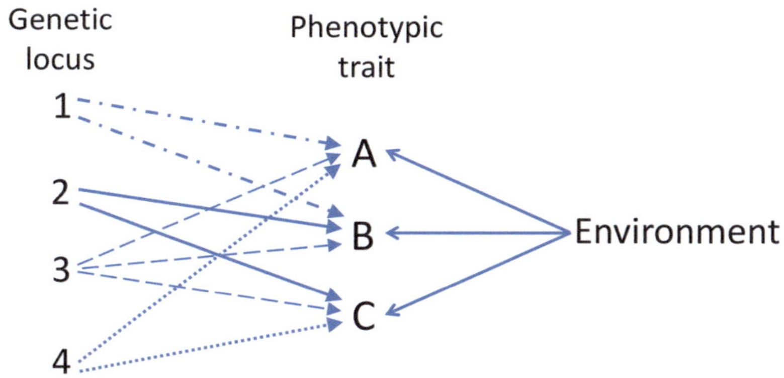


**Fig. 1** The array of influences on continuous variation in biological phenotypes. Morphometric variation in phenotypic traits is polygenic or "multifactorial" (i.e., influenced by more than one allele). Genetic influences on trait form also tend to be pleiotropic (i.e., individual alleles influence more than one trait simultaneously). In addition, phenotypic traits are influenced by "environmental" factors, which can include any nonhereditary source of variation (Modified and redrawn after Conner and Hartl 2004: 156)

multiple genetic loci. That is, most traits observed in organisms (especially quantitative traits) are produced by the aggregate effects of two or more genes, leading them to be described as "multifactorial" or "polygenic" features (Mayr 1982: 791–792). Indeed, many genetic loci may simultaneously affect the form of several different traits, often referred to as "pleiotropic" gene action (Fig. 1).

Reconciling these issues alongside the problem of environmental (i.e., nonhereditary) influences on variation in phenotypic traits (Fig. 1) was approached by the development of what—from the vantage point of privileged hindsight—perhaps seems a simple model. When looking at quantitative variation in a phenotypic trait across a collection ("population") of individuals, this model expresses the total variance ($V_P$) observed as:

$$V_{\text{Phenotype}} = V_{\text{Genetic}} + V_{\text{Environment}}$$

where $V_G$ is the *proportion* of phenotypic variance within the population controlled by genetic factors and $V_E$ is the *proportion* of the variance caused by environmental factors (see, e.g., Falconer 1960). Hence, the phenotypic variance of a trait (as seen across a number of individuals) is the sum of a number of different components that each account for a specific proportion of the total variance. Mutation (i.e., imperfect replication of genetic information) adds to the value of $V_G$ and potentially, therefore, to the total value of $V_P$. An error term ($V_{Err}$) can also be added to this basic formula (sometimes referred to as "residual variance"), which in effect includes everything not explicitly labeled in the major components of the model.

One of the advantages of this modeling framework is that its major components can be further broken down into smaller subcomponents of factors. For example, the proportion of trait variance determined by heritable genetic factors (i.e., $V_G$) can be subdivided into genetic factors caused by additive effects ($V_{Additive}$), dominance effects ($V_{Dominance}$), and interaction effects ($V_{Interaction}$) so that:

$$V_G = V_A + V_D + V_I$$

The definition of these different components does not concern us here (see Lycett and von Cramon-Taubadel 2015); it is sufficient merely to note that major components of the primary formula can be broken down into smaller subconstituent factors, thus providing a flexible framework.

In sum, quantitative genetics explicitly takes account of the *materiality* of the data it is using to study evolution, with the built-in recognition that it is looking at a physical *expression* of an underlying information system that evolves via a process of descent with modification. Importantly, it does so while also acknowledging (and indeed expressly accounting for) the fact that not all of the measurable variation in those physical traits is caused by heritable information. Indeed, as Lycett and von Cramon-Taubadel (2015) have recently noted, one important aspect of the quantitative genetic approach is that it reconciles particular lines of research that might otherwise seem in conflict. For example, it is legitimate to consider questions relating to the environmental (e.g., activity or dietary effects) and developmental (i.e., ontogenetic) effects of variability in a skeletal dataset, since aspects of data variation may well reveal important insights into these factors of variability (e.g., Hoppa and Vaupel 2002; Shaw and Stock 2009; von Cramon-Taubadel 2011). However, this does not negate the fact that examining variability in the same set of data can also indicate issues relating to evolutionary history, once the scale and nature of the questions considered are shifted (e.g., Roseman and Weaver 2007; Betti et al. 2012; von Cramon-Taubadel 2014).

There are many analogies between the problems that required the development of quantitative genetic approaches to phenotypic traits and certain difficulties currently facing the evolutionary analysis of artifactual traits (Lycett and von Cramon-Taubadel 2015). Consider, for example, pottery attributes such as rim shape, neck shape, or handle shape (Fig. 2). On the one hand, several cultural (i.e., socially learnable) factors may simultaneously have an influence on quantifiable aspects of variation in one, or more, attributes of such artifacts, much the same as individual alleles might have an effect on an array of phenotypic variables (Fig. 1). These cultural variants could, for example, include whether the pots are entirely made by hand or wheel-thrown, the method of tempering, differences in the mechanics of firing, etc. (e.g., Orton et al. 1993). The key point here is that many quantitative traits of artifacts segregate across several different heritable causal factors, just as in the case of multifactorial quantitative traits examined in biology. Yet also, some aspects of attribute variability may result from the type of clay to which these socially learned factors are applied—in other words, the "environment" in which the inherited components operate may influence variability in observable attributes too (Fig. 2). Hence, observable archaeological traits will invariably be influenced by multiple unobservable and "pleiotropically" operating cultural elements, while also simultaneously being influenced by "environmental" effects (Fig. 2).

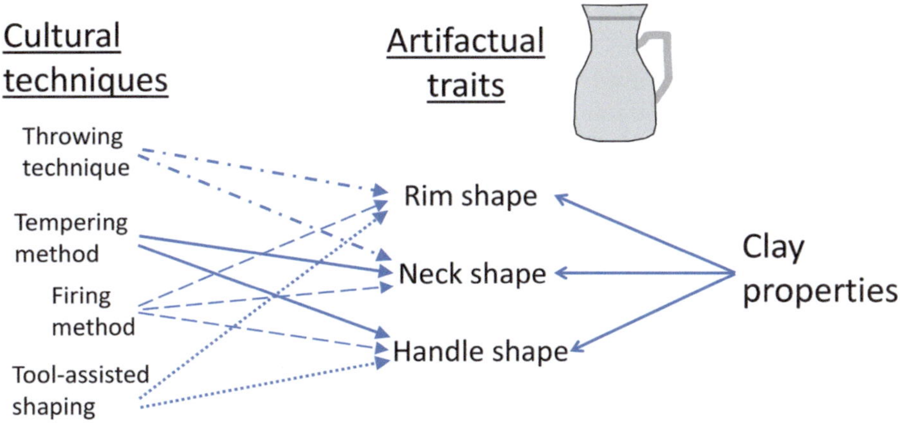


**Fig. 2** Artifactual traits are the product of multiple socially learned input factors. Illustrated here is the idea that variation in many traits of artifacts will be "multifactorial" (i.e., simultaneously influenced by more than one behavioral input variable), and individual behavioral factors (e.g., method of firing pots) may influence more than one trait at a time (analogous to pleiotropic gene action). In addition, artifactual traits will also be influenced by "environmental" variation, which can include any nonhereditary source of variation, such as raw material factors (here illustrated by clay properties). Variation in artifactual traits is, therefore, influenced by an array of (heritable) behavioral factors as well as nonhereditary factors

In the light of a quantitative genetic framework, however, we can model the total variation in an attribute observed across a set of artifacts ($V_{AS}$) as:

$$V_{\text{ArtifactSet}} = V_{\text{Culture}} + V_{\text{RawMaterial}}$$

where $V_C$ is the proportion of total variance measured in an attribute controlled by cultural factors and $V_{RM}$ is the proportion of the variance in that attribute caused by raw material factors (e.g., clay type). It should be noted that factors responsible for $V_C$ constitute *anything* that is inherited by individuals (or is at least potentially heritable) via any mechanism of social learning, consciously or otherwise, such that it comes to propagate and influence the manufacture and final form of artifacts. This would certainly include particular means ("techniques") of making certain classes of artifacts (Fig. 2), but it could also include more abstract concepts about how an artifact "should" look (e.g., based on how other similar items look). Moreover, just as new mutations occur during the transmission of genetic material, social transmission is rarely perfect, such that "copying errors" in all of these socially learned factors may lead to the introduction of novel variants (Eerkens and Lipo 2005; Kempe et al. 2012; Schillinger et al. 2014a, b). This latter factor will increase the proportion of total variance represented by $V_C$ and, in turn, the measurable variability of the attribute in the total artifact set observed (i.e., $V_{AS}$).

Of course, choice of raw material sources might also be culturally guided (i.e., socially learned) rather than the result of random factors. Hypothetically, for instance, imagine a situation where a community discovers and explores the possibilities of ceramic production for the first time. Initially, they choose from a wide variety of clay sources of varying properties and are guided in their choices more by proximity and availability than they are by their particular properties. Over time, however, particular sources of clay are favored because of the fact that the pots function better, and clay is thus chosen in a far less random fashion because of culturally inherited biases. In this sense, "clay choice" is now a cultural variable and so subsumed as part of the information system associated with ceramic production. However, this does not mean that clay has no effect on the traits of pots produced by that community, and so $V_{RM}$ refers to this residual source of variation, just as environmental effects in biology refer to the effects of factors overlaying those caused by heritable (i.e., genetic) information.

Just as in the biological case, this framework facilitates flexibility such that additional factors, or subsets of factors, might be incorporated. For instance, in the case of lithic artifacts, an important factor to consider is their reductive process of manufacture and maintenance (Schillinger et al. 2014a). Several authors have suggested analogies between the reduction of stone tools during their manufacture and the developmental (i.e., ontogenetic) phases of an organism's life (e.g., Iovita 2010; Lycett 2010). Moreover, in the case of lithic artifacts such as flake tools (e.g., Frison 1968) or projectile points (e.g., Flenniken and Raymond 1986), resharpening and repair as a result of usage may introduce further aspects of variability in a manner analogous to "aging" (Shott 2010: 275) or "senescence" (Lycett 2010: 227) in biological phenotypic traits. Lycett and von Cramon-Taubadel (2015) have shown that for lithic artifacts, these matters can be incorporated by extending the main formula such that:

$$V_{\text{ArtifactSet}} = V_{\text{Culture}} + V_{\text{RawMaterial}} + V_{\text{Reduction}}$$

where $V_{\text{Reduction}}$ ($V_R$) is the proportion of an attribute's total variation that can be attributed to reductive factors. This may also be subdivided into reductive factors associated with the production of a tool leading up to its first usage ($V_{\text{OntogenyReduction}}$) plus any reduction associated with retouch or resharpening ($V_{\text{SenescenceReduction}}$), so that:

$$V_{\text{Reduction}} = V_{OR} + V_{SR}$$

Just as in the biological case, a key point to note here is the reconciliatory effect this framework can have on several different lines of research looking at patterns of artifactual variability, which might otherwise seem in conflict. For instance, just because a series of stone projectile points may have aspects of their variation influenced by factors of raw material, or because of resharpening effects, does not negate the fact that statistically meaningful patterns of variation resulting from important evolutionary factors are *also* detectable in those artifacts and indeed allow evolutionary questions to be addressed (Lycett and von Cramon-Taubadel 2015). In other words, in the context of a quantitative genetics framework, viable questions relating to all of these different sources of variation may be validly addressed—they are not mutually exclusive.

## 3 The Importance of "Heritability" and the Capacity for Evolutionary Change: A Quantitative Discussion

Archaeologists have long understood that multiple different factors influence variation in the traits of artifacts that they observe (e.g., Schiffer 1976). This same insight and truth may, however, have hindered the degree to which evolutionary approaches to artifactual traits are accepted, since if a trait's variation is influenced by raw material, how can evolutionary forces (e.g., drift or selection) operate? As noted earlier, the development of a quantitative genetic framework was essential in evolutionary biology in order to more precisely connect statistical variations in the physical traits of plants and animals, to evolutionary processes such as selection and drift given increased recognition of the complexity of factors affecting their variation. This was especially the case when it came to understanding and describing precisely how various selective and stochastic forces (i.e., drift) are linked, *mechanistically*, to subtle (but observable) variations in the different traits of physical forms influenced by both heritable and nonheritable factors. Evolution is undoubtedly a process of "descent with modification" in which the three factors of variation, inheritance of variation, and its differential replication are both the necessary and sufficient factors for its operation. Nevertheless, however, much of this may be correct, in the strictest of terms, a bird's "wingspan" is not inherited by its offspring, nor does one pot "inherit" the shape of its base from another pot; both of these physical features are the material expression of information that is inherited within an evolving information system. One of the important implications of quantitative genetics is that the framework can be used to establish a quantitative concept of "heritability" in light of the fact that not all variability is heritable (Roff 1997; Lynch and Walsh 1998; Conner and Hartl 2004). The parameter of "heritability" is essential in order to understand how selection and drift relates to evolutionary patterns even in the face of nonhereditary sources of variation. Indeed, it is the key to understanding, and explaining, evolution given the complexities of the data under these circumstances.

In order to quantify the potential that factors such as selection (of any form) and/or drift *might* have on the variance and mean value of a trait measured across time or space, it is necessary to have a quantitative estimate of the parameter "heritability" (see, e.g., Falconer and Mackay 1996). In terms of a metric biological phenotypic trait (e.g., "length" of a skull, or beak "shape"), the heritability of the trait (notated as $H^2$) is the proportion of the total phenotypic variance across the population (i.e., $V_P$), expressed as a ratio of the variance attributable to genetic factors (i.e., $V_G$). Hence, simply:

$$H^2 = \frac{V_G}{V_P}$$

The use of the squared symbol here is a reminder that the parameter of heritability is based on the descriptive statistic of variance (i.e., the standard deviation squared) in terms of the two variables used to compute it. Because the denominator ($V_P$) of this fraction is always inevitably larger than the numerator ($V_G$), computed heritability values always range between 0 and 1. (Please note that due to complications that do not concern us here, the parameter described above is strictly referred to as "broad sense" heritability, which is often contrasted with "narrow sense" heritability, or $h^2$ in biological usage.)

In quantitative genetic terms, therefore, the heritability of a metric trait is the degree to which variation in that trait (measured across a group of individuals) is determined by genetic factors, expressed as a *ratio* of the degree to which variation in that trait is also determined by additional (i.e., nonheritable) factors. Because total phenotypic variance ($V_P$) is the sum of both genetic and environmental components (i.e., $V_G + V_E$), one of the main factors affecting the value of $H^2$ is the extent to which variation in the trait is determined by nonheritable (i.e., "environmental") factors. In cases where environment is having a relatively large effect, the computed value of $H^2$ will decrease. If, hypothetically, $H^2 = 0$, then there is no heritable source of variation in that trait, and so evolutionary forces (e.g., selection or drift) cannot influence changes in the mean value and/or variance. In cases where $H^2$ has a value approaching 1, selection or drift has the greatest potential to produce change in the mean and/or variance values of that trait in subsequent generations. Understanding this ratio is, therefore, fundamental to understanding the relationship between how things look in one generation, compared to how they might look in the next generation once drift and/or selection has done its work. Mathematically, this is expressed by what is referred to as "breeder's equation" (see, e.g., Conner and Hartl 2004).

Breeder's equation calculates what is termed the "response to selection" (*R*), using the "selection differential" (*S*), which is simply the difference between the mean value of the entire population and the mean of those individuals who are not subject to negative selection. It is then computed as:

$$R = H^2 S.$$

The "response to selection" (*R*) is, therefore, a value that describes the magnitude of evolutionary change in the mean value of a trait, given the heritability of the trait, and the potency of selection (or random sorting) on the parent generation. The computed value obviously increases when either the strength of selection (*S*) increases or the heritability ($H^2$) increases. As long as both values are not zero, changes in the mean value of the trait will occur.

It is important to note that while breeder's equation predicts the magnitude of change in the mean value of a trait, the variance of the trait may not *necessarily* change in the face of selection (see, e.g., Crow 1986: 121–122). This is partly because some of the variation in the trait is, of course, invariably controlled by "environmental" factors, and these can (potentially) affect the variance of the subsequent generation to the same extent. A further reason is that if the trait is multifactorial, and so is under the simultaneous influence of several different genetic loci (i.e., the trait is "polygenic" as in Fig. 1), then the effects of selection may only diminish the frequency of particular alleles, with the others still contributing to the overall variability of the trait, as before. In other words, selection (or stochastic sorting) may increase the relative contribution that some genetic loci make at the expense of others. A third factor to consider is that genetic inheritance is not perfect; that is, mutation may add new variation around the mean of the trait. In combination, these three factors may, therefore, ensure that while the mean value of a trait changes due to selection, this will not necessarily be reflected in changes of population variance around the mean. As we will see below, in the light of a quantitative genetics approach, these factors need also to be taken into account when conceiving of the relationship between artifacts, evolution, and the effects of selective and/or random sorting on artifactual attributes.

## *3.1 Thinking about "Heritability" Archaeologically: A "Quantitative Genetic" Approach*

Given the forgoing, let us consider how these matters can inform on the issue of artifactual variation using the "quantitative genetic" framework outlined. Figure 3 shows a hypothetical set of 10 Acheulean handaxes that vary in quantitative terms. In the computations that follow, the size variable of "length" is used to illustrate the principles, but the same computations could be undertaken for any quantitative (morphometric) character that varies continuously, including size-adjusted shape variables. Using the handaxe length values shown in Fig. 3, we can compute the mean length of all 10 handaxes as 12.05 cm and the variance ($\sigma^2$) as 6. Given the

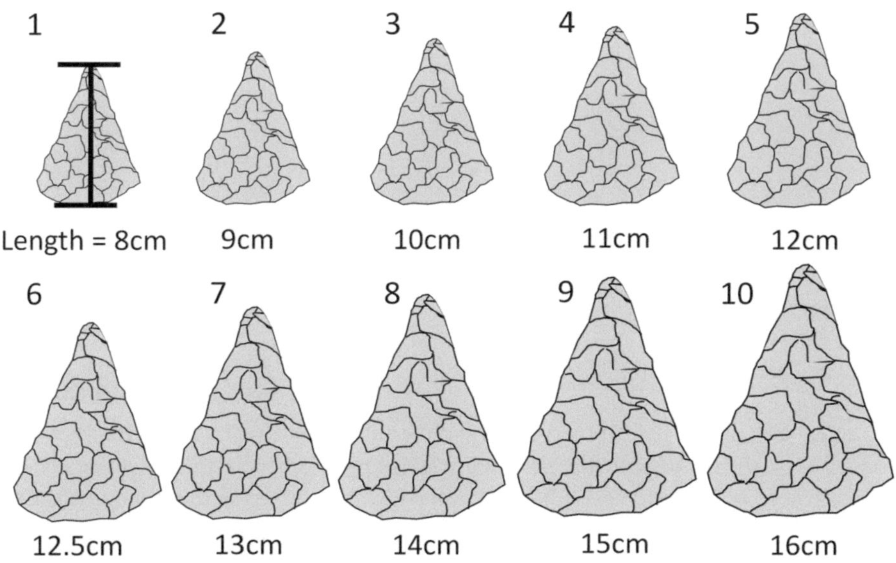


**Fig. 3** A series of 10 hypothetical handaxes that vary morphometrically in terms of the variable "length"

framework outlined earlier, we can therefore compute heritability ($H^2$) for this set of handaxes as:

$$H^2 = \frac{V_C}{V_{AS}}$$

where, as before, $V_{AS}$ is the total variance of the trait across the artifact set and $V_C$ is the proportion of that variance that can be attributed to cultural factors. Let us assume that raw material ($V_{RM}$) accounts for 30 % of the total variance; it should be noted that this percentage is not entirely inconsistent with real archaeological data given the accuracy with which handaxes can be correctly assigned to their localities in multivariate analyses (e.g., Lycett and Gowlett 2008; Lycett and von Cramon-Taubadel 2015). Given, therefore, that $V_{AS}=6$, we can determine that $V_C=4.2$ (i.e., 70 % of total variance). Given these values, we can compute heritability ($H^2$) for the trait of "length" in this set of handaxes as:

$$H^2 = \frac{V_C}{V_{AS}} = \frac{4.2}{6.0} = 0.7$$

Given this heritability value, let us assume, hypothetically, that negative selection is operating against the five smallest handaxes under the scenario that larger handaxes perform more efficiently as tools. Given that the five largest handaxes are, therefore, preferentially copied, we can compute the "selection differential" ($S$) as the difference between the mean value of the entire population and the mean of those individuals who are not subject to negative selection.

Given these variables, we can compute the response to selection ($R$) as:

$$\begin{aligned} R &= H^2 S \\ &= 0.7 \times (14.1 - 12.05) \\ &= 0.7 \times 2.05 \\ &= 1.435 \end{aligned}$$

Put another way, we can predict that the mean length in handaxes would increase in the next "generation" by ~1.44 cm over that exhibited in the previous generation. If this were repeated over just five generations, the mean length of handaxes would have increased by 7.2 cm, that is, from 12.05 to 19.25 cm (62 %). This is despite the fact that not all of the variation is controlled by behavioral factors and that 30 % of the total variance is determined by raw material factors. Just as in the biological case, it is important to note that the change in mean values will not, however, *necessarily* lead to a change in overall variance levels. Analogous to the biological case, this is because "environment" (in this case raw material) will continue to have an effect. Also, because such an instance of selection may only have influenced the relative frequency of a small number of the cultural factors leading to the production of handaxes, all other remaining variables will continue to influence total variation. Moreover, just as genetic mutation may add variation "back in" to the offspring population in the case of biology, so the existence of copying errors during social transmission events will potentially elevate measurable levels of trait variation (see, e.g., Eerkens and Lipo 2005; Kempe et al. 2012; Schillinger et al. 2014a, b).

Let us now consider the potential effect of stochastic sorting leading to "drift" of a variable such as length under such conditions. Keeping all other parameters the same, let us sample five handaxes from the original set of ten (Fig. 3) at random. Using a random number generator (http://www.randomizer.org), the handaxes *negatively* selected were 1, 3, 4, 8, and 9. The mean length value for the remainder is 12.5 cm. Given this, we can now compute the response to selection ($R$) as:

$$\begin{aligned} R &= H^2 S \\ &= 0.7 \times (12.5 - 12.05) \\ &= 0.7 \times 0.45 \\ &= 0.315 \end{aligned}$$

Hence, in this case, the mean value would increase by ~0.32 cm in the next generation, just by drift alone. While this amount may seem trivial, the effect could amount to substantial change over time if similar sampling effects were repeated. Again, it should be noted that this takes place despite the fact that a substantial proportion (30 %) of artifactual variance cannot be controlled by cultural transmission factors.

What these exercises illustrate is the fundamental role that a quantitative concept of heritability must play in bringing about measurable change in artifact traditions, even when not all of the variation in a trait is heritable, and when we are looking at a proxy variable of the evolutionary process. In archaeological terms, therefore, the heritability of a metric artifactual trait is the degree to which variation in a trait (measured across a set of items) is determined by socially learnable behavioral factors, expressed as a ratio of the degree to which variation in that trait is also determined by additional (i.e., nonheritable) factors. Just as much as in the biological case, understanding this ratio is fundamental to understanding the relationship between how things (i.e., artifacts) look in one generation compared to how they might look in the next generation once selection and/or drift has done its work.

These exercises also illustrate that as long as some proportion of the variance in an artifactual trait is caused by socially transmitted factors, then selection and drift have the capacity to bring about measureable change. If $H^2$ values are relatively low, such change may occur slower than in instances where $H^2$ values are higher; but wherever *some* proportion of the variance is under the control of socially transmitted parameters, then, there is potential for measurable change brought about by evolutionarily historical factors. This is important when we consider the diversity of selective biases that can potentially operate on artifactual traits, including conformity bias, prestige bias, frequency-dependent bias (e.g., rarity), performance (or functional) bias, aesthetic bias, as well as natural selection (see, e.g., Boyd and Richerson 1985; Henrich and McElreath 2003; Lycett 2008; Mesoudi 2011 for discussion of various categories of selective biases that may operate in cultural terms). These biases are joined, of course, by the ever-present possibility of stochastic sampling caused by chance factors. Given this multiplicity of factors, the metric aspects of artifactual traits are unlikely to be represented equally during the course of repeated cultural transmission events. With the simple requirement that just *some* proportion of the measurable variance be under the control of socially learnable factors, and *some* aspect of unequal sampling of that variance occurring, the potency for measurable *evolutionary* change can be appreciated given the quantitative genetic approach outlined.

Of course, much work will be required to obtain more accurate estimates of the key parameter of $V_{RM}$ in archaeological assemblages, such that more accurate determinations of $H^2$ can be computed in a range of artifacts. A "quantitative genetic" perspective can help provide the statistical framework, but it cannot supply the raw data; for this we must turn to the—thankfully rich—potential of the archaeological record.

# 4 Conclusions

The major strength of the archaeological record is its potential to reveal statistical patterns over temporal and spatial scales (Clarke 1968). Linking the raw statistical data of artifactual variation to evolutionary processes is, however, by no means straightforward, especially when some of that variation is caused by factors such as raw material. An early scholar of biological phenotypic evolution stated that it "cannot be too strongly urged that the problem of animal evolution is essentially a statistical problem" (Weldon 1893: 329). It could equally be stressed, however, that this is a *complex* statistical problem. This complexity arises directly because of the circuitous relationship between the evolutionary process and its measurable physical remnants (which comprise the basic data) as well as the fact that variation in those data is also caused by additional factors such as environment. Quantitative genetics helped resolve these difficulties by more explicitly quantifying the dynamics of these factors and their interactions. It is one of the central arguments of this chapter that a similar framework can help to resolve similar problems in the case of linking variations in the physical traits of artifacts to evolutionary dynamics such as stochastic factors (drift) and selection biases.

The benefits of this approach are, therefore, twofold: the first is practical, and the second is theoretical. The practical benefit is that it provides a statistical basis for studying artifactual variation and evolutionary patterning, framed expressly in *material* terms. As has been argued here, study of cultural evolution via the archaeological record requires models that are framed—as far as is possible—in directly material terms, taking account of the range of factors affecting that materiality, especially when these factors cannot be directly observed, or even necessarily inferred, solely via measurements of artifactual variation. The approach moves beyond the primary elements of "descent with modification" to express, in more precise terms, the statistical *mechanics* of the relationship between selective biases, stochastic factors, and the patterns they create in measureable data. As an outgrowth, the theoretical benefit is that it will help to refocus the way we frame and explain *what* is being studied and *how* it is being used to examine the evolution of the underlying information system—something which is entirely unobservable.

This is not to say, of course, that the sentences above solve all of the problems raised or indeed that this framework is yet complete. In biology, the field of quantitative genetics required the application of many great minds and a great deal of hard work to flesh out the details, to measure and compute the effects of the models, to simulate their effects, and to test the various elements with empirical data. Thankfully, however, in the case of archaeology, the key elements required are already being rigorously explored by the current generation of evolutionarily minded archaeologists. A quantitative genetics approach, however, will allow new connections to be made between the results of such studies, patterns of artifactual variation, and the evolution of the underlying information system. As a result, the complex interplay between cultural evolution and its material proxy (i.e., spatiotemporal variation in artifacts) may be unraveled more clearly.

**Acknowledgements** I am grateful to Larissa Mendoza Straffon for her invitation to contribute to this volume. My application of quantitative genetic principles to artifactual data has developed as a direct result of collaborations and innumerable conversations with Noreen von Cramon-Taubadel; I am indebted for the essential role her input has played.

## References

Bentley, R. A., & Shennan, S. J. (2003). Cultural transmission and stochastic network growth. *American Antiquity, 68*(3), 459–485.

Bentley, R. A., Hahn, M. W., & Shennan, S. (2004). Random drift and culture change. *Proceedings of the Royal Society of London B, 271*, 1443–1450.

Betti, L., von Cramon-Taubadel, N., & Lycett, S. J. (2012). Human pelvis and long bones reveal differential preservation of ancient population history and migration out of Africa. *Human Biology, 84*(2), 139–152.

Bettinger, R. L., & Eerkens, J. (1999). Point typologies, cultural transmission, and the spread of bow-and-arrow technology in the prehistoric Great Basin. *American Antiquity, 64*(2), 231–242.

Bowler, P. J. (2003). *Evolution: The history of an idea*. Berkeley: University of California Press.

Boyd, R., & Richerson, P. (1985). *Culture and the evolutionary process*. Chicago: University of Chicago Press.

Brantingham, P. J., & Perreault, C. (2010). Detecting the effects of selective and stochastic forces in archaeological assemblages. *Journal of Archaeological Science, 37*, 3211–3225.

Brew, I.O. (1946). *Archaeology of Alkali Ridge, Southeastern Utah*. Peabody Museum of American Archaeology and Ethnology Papers 21.

Buchanan, B., & Collard, M. (2007). Investigating the peopling of North America through cladistic analyses of Early Paleoindian projectile points. *Journal of Anthropological Archaeology, 26*(3), 366–393.

Buchanan, B., & Collard, M. (2008). Phenetics, cladistics, and the search for the Alaskan ancestors of the Paleoindians: A reassessment of relationships among the Clovis, Nenana, and Denali archaeological complexes. *Journal of Archaeological Science, 35*, 1683–1694.
Buchanan, B., & Hamilton, M. (2009). A formal test of the origin of variation in North American early Paleoindian projectile points. *American Antiquity, 74*, 279–298.
Byrne, R. W., & Russon, A. E. (1998). Learning by imitation: a hierarchical approach. *Behavioral and Brain Sciences, 21*, 667–721.
Cavalli-Sforza, L. L., & Feldman, M. W. (1981). *Cultural transmission and evolution: A quantitative approach*. Princeton: Princeton University Press.
Clarke, D. L. (1968). *Analytical archaeology*. London: Methuen.
Cochrane, E. E., & Lipo., C. P. (2010). Phylogenetic analyses of Lapita decoration do not support branching evolution or regional population structure during colonization of remote Oceania. *Philosophical Transactions of the Royal Society of London. Series B, Biological Sciences, 365*(1559), 3889–3902.
Cochrane, E. E., Rieth, T. M., & Dickinson, W. R. (2013). Plainware ceramics from Sāmoa: Insights into ceramic chronology, cultural transmission, and selection among colonizing populations. *Journal of Anthropological Archaeology, 32*(4), 499–510.
Conner, J. K., & Hartl, D. L. (2004). *A primer of ecological genetics*. Sunderland: Sinauer.
Crow, J. F. (1986). *Basic concepts in population, quantitative, and evolutionary genetics*. New York: Freeman.
Darwin, C. (1859). *On the origin of species*. London: John Murray.
Dawkins, R. (1989). *The selfish gene*. Oxford: Oxford University Press.
Dunnell, R. C. (1978). Style and function: a fundamental dichotomy. *American Antiquity, 43*(2), 192–202.
Dunnell, R. C. (1980). Evolutionary theory and archaeology. In M. B. Schiffer (Ed.), *Advances in archaeological method and theory* (Vol. 3, pp. 35–99). New York: Academic Press.
Eerkens, J. W., & Lipo, C. P. (2005). Cultural transmission, copying errors, and the generation of variation in material culture and the archaeological record. *Journal of Anthropological Archaeology, 24,* 316–334.
Eerkens, J. W., & Lipo, C. P. (2007). Cultural transmission theory and the archaeological record: context to understanding variation and temporal changes in material culture. *Journal of Archaeological Research, 15*, 239–274.
Ellen, R., Lycett, S. J., & Johns, S. E. (2013). *Understanding cultural transmission in anthropology: A critical synthesis*. New York: Berghahn.
Falconer, D. S. (1960). *Introduction to quantitative genetics*. New York: Ronald.
Falconer, D. S., & Mackay, T. F. C. (1996). *Introduction to quantitative genetics* (4th ed.), Harlow: Pearson.
Fisher, R. A. (1918). The correlations between relatives on the supposition of Mendelian inheritance. *Trans Roy Soc Edinburgh, 52*(2), 399–433.
Flenniken, J. J., & Raymond, A. W. (1986). Morphological projectile point typology: Replication experimentation and technological analysis. *American Antiquity, 51*, 603–614.
Frison, G. C. (1968). A functional analysis of certain chipped stone tools. *American Antiquity, 33*:149–155.
Hamilton. M. J., & Buchanan, B. (2009). The accumulation of stochastic copying errors causes drift in culturally transmitted technologies: quantifying Clovis evolutionary dynamics. *Journal of Anthropological Archaeology, 28*, 55–69.
Henrich, J., & McElreath, R. (2003). The evolution of cultural evolution. *Evolutionary Anthropology, 12*, 123–135.
Hoppa, R. D., & Vaupel, J. W. (Eds.). (2002) *Paleodemography: Age distributions from skeletal samples*. Cambridge: Cambridge University Press.
Hurt, T. D., & Rakita G. F. M. (2001). *Style and function: Conceptual issues in evolutionary archaeology*. Westport: Bergin & Garvey.
Iovita, R. P. (2010). Comparing stone tool resharpening trajectories with the aid of elliptical Fourier analysis. In S. J. Lycett & P. R. Chauhan (Eds.) *New Perspectives on Old Stones: Analytical Approaches to Palaeolithic Technologies* (pp. 235–253). New York: Springer.

Jennings. T. A., & Waters, M. R. (2014). Pre-clovis lithic technology at the Debra L. Friedkin Site, Texas: Comparisons to Clovis through site-level behavior, technological trait-list, and cladistic analyses. *American Antiquity*, *79*(1), 25–44.

Johannsen, W. (1909). *Elemente der exakten Erblichkeitslehre*. Jena: Gustav Fischer.

Jordan, P., & O'Neill, S. (2010). Untangling cultural inheritance: language diversity and long-house architecture on the Pacific northwest coast. *Philosophical Transactions of the Royal Society B: Biological Sciences*, *365*(1559), 3875–3888.

Jordan, P., & Shennan, S. J. (2003). Cultural transmission, language and basketry traditions amongst the California Indians. *Journal of Anthropological Archaeology, 22*, 42–74.

Jordan, P., & Shennan, S. J. (2009). Diversity in hunter–gatherer technological traditions: mapping trajectories of cultural 'descent with modification' in northeast California. *Journal of Anthropological Archaeology*, *28*, 342–365.

Kempe, M., Lycett, S. J., & Mesoudi, A. (2012). An experimental test of the accumulated copying error model of cultural mutation for Acheulean handaxe size. *PLoS ONE*, *7*(11), e48333.

Kohler, T. A., Van Buskirk, S., & Ruscavage-Barz, S. (2004). Vessels and villages: Evidence for conformist transmission in early village aggregations on the Pajarito Plateau, New Mexico. *Journal of Anthropological Archaeology*, *23*, 100–118.

Leonard, R. D., & Jones, G. T. (1987). Elements of an inclusive evolutionary model for archaeology. *Journal of Anthropological Archaeology*, *6*(3), 199–219.

Lipo, C. P., Madsen, M. E., Dunnell, R. C., & Hunt, T. (1997). Population structure, cultural transmission, and frequency seriation. *Journal of Anthropological Archaeology, 16,* 301–333.

Lipo, C. P., O'Brien, M. J., Collard, M., & Shennan, S. (2006). *Mapping our ancestors: Phylogenetic approaches in anthropology and prehistory*. New Brunswick: Aldine Transaction.

Lipo, C. P., Hunt, T. L., & Hundtoft, B. (2010). Stylistic variability of stemmed obsidian tools (*mata'a*), frequency seriation, and the scale of social interaction on Rapa Nui (Easter Island). *Journal of Archaeological Science, 37*(10), 2551–2561.

Lycett, S. J. (2007). Why is there a lack of Mode 3 Levallois technologies in East Asia? A phylogenetic test of the Movius-Schick hypothesis. *Journal of Anthropological Archaeology*, *26*(4), 541–575.

Lycett, S. J. (2008). Acheulean variation and selection: does handaxe symmetry fit neutral expectations? *Journal of Archaeological Science*, *35*(9), 2640–2648.

Lycett, S. J. (2009a). Are victoria west cores 'proto-Levallois'? A phylogenetic assessment. *Journal of Human Evolution*, *56*(2), 175–191.

Lycett, S. J. (2009b). Understanding ancient hominin dispersals using artefactual data: a phylogeographic analysis of Acheulean handaxes. *PLoS ONE, 4*(10/e7404), 1–6.

Lycett, S. J. (2010). Cultural Transmission, genetic models and Palaeolithic variability: Integrative analytical approaches. In S, J. Lycett & P. R. Chauhan (Eds.), *New perspectives on old stones: Analytical approaches to palaeolithic technologies* (pp. 207–234). New York: Springer.

Lycett, S. J. (2011). "Most beautiful and most wonderful": Those endless stone tool forms. *Journal of Evolutionary Psychology*, *9*(2), 143–171.

Lycett, S. J. (2013). Cultural transmission theory and fossil hominin behaviour: A discussion of epistemological and methodological strengths. In R. F. Ellen, S. J. Lycett & S. E. Johns (Eds.), *Understanding cultural transmission in anthropology: A critical synthesis* (pp. 102–130). New York: Berghahn.

Lycett, S. J., Chauhan, P. R. (2010a) Analytical approaches to palaeolithic technologies: An introduction. In S. J. Lycett & P. R. Chauhan (Eds.), *New perspectives on old stones: Analytical approaches to palaeolithic technologies* (pp. 1–22). New York: Springer.

Lycett, S. J., & Chauhan, P. R. (Eds.), (2010b) *New perspectives on old stones: Analytical approaches to palaeolithic technologies*. New York: Springer.

Lycett, S. J., & Gowlett, J. A. J. (2008). On questions surrounding the Acheulean 'tradition'. *World Archaeology, 40*(3), 295–315.

Lycett, S. J., & von Cramon-Taubadel, N. (2008). Acheulean variability and hominin dispersals: A model-bound approach. *Journal of Archaeological Science*, *35*(3), 553–562.

Lycett, S. J., & von Cramon-Taubadel, N. (2015). Toward a "quantitative genetic" approach to lithic variation. *Journal of Archaeological Method and Theory*. http://dx.doi.org/10.1007/s10816-013-9200-9.

Lyman, R. L., VanPool, T. L., & O'Brien, M. J. (2008). Variation in North American dart points and arrow points when one or both are present. *Journal of Archaeological Science, 35*(10), 2805–2812.

Lynch, M., & Walsh, B. (1998). *Genetics and analysis of quantitative traits*. Sinauer: Sunderland.

Mace, R, Holden, C.J., & Shennan, S. (Eds.), (2005). *The evolution of cultural diversity: a phylogenetic approach*. London: UCL Press.

Marwick, B. (2012). A cladistic evaluation of ancient Thai bronze Buddha images: Six tests for a phylogenetic signal in the Griswold collection. In D. Bonatz, A. Reinecke, M. A. Tjoa-Bonatz (Eds.), *Connecting empires* (pp. 159–176). Singapore: National University of Singapore Press.

Mather, K. (1943). Polygenic inheritance and natural selection. *Biol Rev, 18*(1), 32–64.

Matthews, L. J., Tehrani, J. J., Jordan, F. M., Collard, M., & Nunn, C. L. (2011). Testing for divergent transmission histories among cultural characters: a study using Bayesian phylogenetic methods and Iranian tribal textile data. *PLoS ONE, 6*(4), e14810.

Mayr, E. (1982). *The growth of biological thought*. Cambridge: Harvard University Press.

Mesoudi, A. (2011). *Cultural evolution: How Darwinian theory can explain human culture & synthesize the social sciences*. Chicago: Chicago University Press.

Mesoudi, A., & O'Brien, M. J. (2008). The cultural transmission of Great Basin projectile point technology I: An experimental simulation. *American Antiquity, 73*(1), 3–28.

Mesoudi, A., Whiten, A., & Laland, K. N. (2004). Is human cultural evolution Darwinian? Evidence reviewed from the perspective of *The Origin of Species*. *Evolution. 58*(1):1–11.

O'Brien, M. J. (1996a). *Evolutionary archaeology, preface* (pp. xiii-xv). Salt Lake City: University of Utah Press.

O'Brien, M. J. (Ed.), (1996b) *Evolutionary archaeology*. Salt Lake City: University of Utah Press

O'Brien, M. J. (2008). *Cultural transmission and archaeology*. Washington, DC: Society for American Archaeology.

O'Brien, M., & Holland, T. D. (1990). Variation, selection, and the archaeological record. In: M. B. Schiffer (Ed.), *Archaeological method and theory* (vol 2, pp 31–79). Tucson: The University of Arizona Press.

O'Brien, M. J., & Lyman, R. L. (2000). *Applying evolutionary archaeology: A systematic approach*. New York: Kluwer Academic/Plenum.

O'Brien, M. J., & Lyman, R. L. (2003). *Cladistics and archaeology*. Salt Lake City: University of Utah Press.

O'Brien, M. J., & Shennan, S. J. (2010). *Innovation in cultural systems: Contributions from evolutionary anthropology*. Cambridge: MIT Press.

O'Brien, M. J., & Wilson, H. C. (1988). A paradigmatic shift in the search for the origin of agriculture. *American Anthropologist, 90*, 958–965.

O'Brien, M. J., Darwent, J., & Lyman, R. L. (2001) Cladistics is useful for reconstructing archaeological phylogenies: Palaeoindian points from the southeastern United States. *Journal of Archaeological Science, 28*, 1115–1136.

O'Brien, M. J., Lyman, R. L., Mesoudi, A., & Van Pool, T. L. (2010). Cultural traits as units of analysis. *Philosophical Transactions of the Royal Society B, 365*(1559), 3797–3806.

Okumura, M., & Araujo, A. G. M. (2014). Long-term cultural stability in hunter–gatherers: a case study using traditional and geometric morphometric analysis of lithic stemmed bifacial points from Southern Brazil. *Journal of Archaeological Science, 45,* 59–71.

Orton, C., Tyers, P., & Vince, A. (1993). *Pottery in archaeology*. Cambridge: Cambridge University Press.

Provine, W. B. (1971). The origins of theoretical population genetics. Chicago: University of Chicago Press.

Rindos, D. (1989). Undirected variation and the Darwinian explanation of cultural change. In M. B, Schiffer (Ed.), *Archaeological method and theory* (vol 1, pp. 1–45). Tucson: The University of Arizona Press.

Roff, D. A. (1997). *Evolutionary quantitative genetics*. New York: Chapman & Hall.

Roff, D. A. (2007). A centennial celebration for quantitative genetics. *Evolution, 61*(5), 1017–1032.

Rogers, D. S., & Ehrlich, P. R. (2008). Natural selection and cultural rates of change. *Proceedings of the National Academy of Sciences USA*, *105*, 3416–3420.
Rogers, D. S., Feldman, M. W., & Ehrlich, P. R. (2009). Inferring population histories using cultural data. *Proceedings of the Royal Society of London B*, *276*, 3835–3843.
Roseman, C. C., & Weaver, T. D. (2007). Molecules versus morphology? Not for the human cranium. *BioEssays*, *29*, 1185–1188.
Schiffer, M. B. (1976). *Behavioral archeology*. New York: Academic Press
Schillinger, K., Mesoudi, A., & Lycett, S. J. (2014a). Copying error and the cultural evolution of "additive" versus "reductive" material traditions: an experimental assessment. *American Antiquity, 79*(1), 128–143.
Schillinger, K., Mesoudi, A., & Lycett, S. J. (2014b). Considering the role of time budgets on copy-error rates in material culture traditions: An experimental assessment. *PLoS ONE, 9*(5), e97157.
Shaw, C. N., & Stock, J. T. (2009). Intensity, repetitiveness, and directionality of habitual adolescent mobility patterns influence the tibial diaphysis morphology of athletes. *American Journal of Physical Anthropology, 140*, 149–159.
Shennan, S. (2000). Population, culture history, and the dynamics of culture change. *Current Anthropology*, *41*(5), 811–835.
Shennan, S. (2004). Analytical archaeology. In J. Bintliff (Ed.), *A companion to archaeology* (pp. 3–20). Oxford: Blackwell.
Shennan, S. (2006). From cultural history to cultural evolution: an archaeological perspective on social information transmission. In J. C. K. Wells, S. Strickland, K. Laland (Eds.), *Social information transmission and human biology* (pp 173–189). New York: Taylor & Francis.
Shennan, S. (2009). *Pattern and process in cultural evolution*, Los Angeles: University of California Press.
Shennan, S. J., & Wilkinson, J. R. (2001). Ceramic style change and neutral evolution: A case study from Neolithic Europe. *American Antiquity, 66*(4), 577–593.
Shott, M. J. (2010). Stone-tool demography: reduction distributions in North American Paleoindian tools. In S. J. Lycett, & P. R. Chauhan (Eds.), *New perspectives on old stones: Analytical approaches to paleolithic technologies*, (pp 275–293). New York: Springer.
Steele, J., Glatz, C., & Kandler, A. (2010). Ceramic diversity, random copying, and tests for selectivity in ceramic production. *Journal of Archaeological Science, 37*(6), 1348–1358.
Tehrani, J., & Collard, M. (2002). Investigating cultural evolution through biological phylogenetic analyses of Turkmen textiles. *Journal of Anthropological Archaeology*, *21*, 443–463.
Thornton, A., & Raihani, N. J. (2010). Identifying teaching in wild animals. *Learning and Behavior, 38*, 297–309.
Trigger, B. G. (1989). *A history of archaeological thought*. Cambridge: Cambridge University Press.
Van Pool, T. L. (2001). Style, function, and variation: identifying the evolutionary importance of traits in the archaeological record. In T. D. Hurt, G. F. M. Rakita (Eds.), *Style and function: Conceptual issues in evolutionary archaeology* (pp. 119–140), Westport: Bergin & Garvery.
Vaughan, C. D. (2001). A million years of style and function: regional and temporal variation in Acheulean handaxes. In T. D. Hurt & G. F. M. Rakita (Eds.), *Style and function: Conceptual issues in evolutionary archaeology* (pp. 141–163). Connecticut: Bergin & Garvey, Westport.
von Cramon-Taubadel, N. (2011). Global human mandibular variation reflects differences in agricultural and hunter–gatherer subsistence strategies. *Proceedings of the National Academy of Sciences, 108*, 19546–19551.
von Cramon-Taubadel, N. (2014). Evolutionary insights into global patterns of human cranial diversity: Population history, climatic and dietary effects. *Journal of Anthropological Sciences, 92*, 43–77.
Weldon, W. F. R. (1893). On certain correlated variations in *Carcinus moenas*. *Proceedings of the Royal Society*, *54*, 318–329.
Whiten, A., Horner, V., Litchfield, C. A., & Marshall-Pescini, S. (2004). How do apes ape? *Learning & Behavior*, *32*(1), 36–52.
Wright, S. (1931). Evolution in Mendelian populations. *Genetics*, *16*(2), 97–126.

# Part II
# Case Studies and Applications

# Resisting Innovation? Learning, Cultural Evolution and the Potter's Wheel in the Mediterranean Bronze Age

**Carl Knappett**

**Abstract** Although the use of neo-Darwinian models to explain culture change has become quite common in some subfields of archaeology, there remains much resistance within 'interpretive' archaeologies to what is perceived as the simplistic 'biologisation' of culture. Some recent work has sought to build bridges between evolutionary and interpretive archaeologies, with the topic of 'learning' emerging as a useful middle ground between these two standpoints. Yet significant barriers remain to a more thorough integration. Here I identify what appear to be two such barriers: one is the continued commitment in neo-Darwinian approaches to a Cartesian notion of 'information' and the second is the related adherence to the idea of distinct cultural 'traits'. I draw on work in cognitive science and developmental biology that places heavy emphasis on the distributed and contextual nature of learning, such that the uptake of an innovative technology cannot be reduced to a process of information transfer for learning a new trait. A distributed and developmental approach is put into play through a case study tackling the variable regional and temporal adoption of the potter's wheel across the Bronze Age east Mediterranean.

**Keywords** innovation • learning • potter's wheel • scaffolding • cognitive • Bronze Age • Aegean • Mediterranean

In this paper I use an archaeological example to address processes of technological innovation, challenging the validity of most neo-Darwinian biological models for explaining certain important kinds of cultural change. Archaeology can provide useful case studies because of its capacity to identify the spread of artefacts and techniques over broad regions and long time spans. One such case study is the technique of the potter's wheel. For potters to start using the wheel technique entails

---

C. Knappett
Department of Art, University of Toronto, Toronto, ON, Canada
e-mail: carl.knappett@utoronto.ca

 
L. Mendoza Straffon (ed.), *Cultural Phylogenetics: Concepts and Applications in Archaeology*, Interdisciplinary Evolution Research,
DOI 10.1007/978-3-319-25928-4_5

quite a dramatic shift in their motor skills and habits, as compared to hand-building techniques, and requires a major investment in learning. Moreover, the technique can be identified archaeologically through the forming traces left on the pottery itself, which means it can be tracked quite readily in pottery assemblages. The east Mediterranean has very diverse and abundant assemblages of pottery across many sites during the Bronze Age, which is when the wheel technique appears. However, rather than a predictable spread of the technique, with potters adopting the technique once they learn of it, we actually observe a very uneven picture of adoption and resistance over many centuries.

This innovation then provides a challenge to some typical neo-Darwinian approaches to cultural evolutionary change. In such approaches, it seems difficult to get away from the idea that culture consists of units of 'information' of some kind. As Jordan (2014, 2) quite explicitly puts it:

> human technological traditions…consist of information stored in human brains that is then passed on to other individuals through social learning.

Jordan goes on to assert that technological traditions are 'material manifestations of a complex transmission system' (2014, 2). So we have here a conception of culture as bundles of information in the brain—a set of mental states or ideas. This means that the material world is, in effect, epiphenomenal. The focus on learning that Jordan espouses is very timely and does form a useful response to some of the serious criticisms of neo-Darwinian approaches; but it nonetheless reduces learning to an exploration of the various social ways in which humans manage to pass on cultural information from one brain to another. There is little room for the environment as an active participant in the structuring or constituting of cultural knowledge. Ironically, Jordan does acknowledge scaffolding and other developmental tenets in his bridge building exercise with interpretive archaeologies—but this recognition does not stretch to really taking on board the cognitive implications of scaffolding.

A second, related limitation concerns the definition of the cultural 'trait' as a basic unit of technological traditions. Jordan (2014, 5–6) argues that specific design traits can be combined in various ways, and it is the choices of how to combine them that make up material culture traditions. His use of the term 'design grammars' betrays the conception of these combinations as functioning in a modular manner, like language. However, material cultural traits are not created as interchangeable units readily understood according to formal convention. Generally speaking, there does not seem to exist a language of material culture traits such that, within pottery making, for example, one can just swap out coil building for wheel throwing as if they are interchangeable modules. This is what Jordan implies when he says that 'the choices made at different stages in the production of complex technologies like basketry, skis, or tailored clothing can be defined as "cultural traits", with the presence and the absence of such traits documented across different artefacts' (Jordan 2014, 14). Underlying this standpoint are the same kind of internalist assumptions that guide the neo-Darwinian approach to learning: traits are primarily cultural ideas, unaffected by their externalisation in material form.

I have quoted Jordan as exemplary here, but this is simply because he has expressed some of these issues most clearly in his important recent contribution. These conceptions of information and traits seem quite widely shared across many evolutionary approaches. It is revealing, I think, that in another recent attempt to bridge the wide gap between evolutionary and interpretive archaeologies, the problem of these cognitivist assumptions is not raised at all (Gardner and Cochrane 2011). However, in one contribution sympathetic to (yet also critical of) evolutionary aims, Roux (2013) does recognise the contribution of both cognitive factors and social context in the learning of a 'trait' (see also her work with Blandine Bril—Roux and Bril 2005). As Roux puts it, evolutionary models rely on 'the hypothesis that contacts between people are necessary and sufficient for social learning to occur' (Roux 2013, 313).[1] Roux challenges this basic assertion of social Darwinian approaches, noting that there are many documented instances in which people may 'learn' about a new trait or technique, but nevertheless choose not to adopt it. The decision to adopt, she stresses, depends not just on the existence of contact but on the *nature* of that contact. Furthermore, in those instances where traits actually do spread, Roux argues that there may be quite different processes at work, even with the very same technology. And interestingly enough, it is the example of the potter's wheel that Roux uses to demonstrate her point. She has done extensive research on the uptake of this technique across the Near East. So, for example, in the northern Levant, the process of adoption of the potter's wheel is quite distinct from the process of adoption in the southern Levant. For the former case, she describes a process of 'cultural diffusion', with the wheel technique slowly being adopted by local groups through gradual copying (Roux 2013, 320–24). For the latter, however, the change is much more sudden, and the technique appears to be spread through population mobility and the influx of new groups into the area from the northern Levant; this she labels 'demic diffusion'.

In both cases, however, Roux is describing scenarios in which the potter's wheel *does* see widespread uptake. In an earlier paper, Roux considers examples where the wheel does not catch on—notably during earlier periods in the southern Levant. Here the innovation of wheel fashioning[2] is first seen in the Chalcolithic period (c. 4000–3500 BC), used principally for making ceremonial bowls. Yet, with the start of Early Bronze (EB) I, c. 3500 BC, the wheel method disappears amid significant settlement contraction and other cultural changes (Roux 2010, 227). So the technique never fully establishes itself. Such scenarios place even more strain on the

[1] Wengrow (2011) also offers a challenge to the neo-Darwinian assumption of contact being a sufficient condition for the diffusion of traits. He critiques Boyer's notion that counter-intuitive traits (like monster images) will simply spread as if by contagion—a kind of 'epidemiology' of culture model.

[2] The 'wheel-fashioning' technique involves the application of rotative kinetic energy to a coil-built body. It differs from 'wheel throwing', which entails the use of rotative kinetic energy to draw up the clay body 'from the hump' and no involvement of coils. Both methods can be grouped together under the broader term of 'wheel made'.

plausibility of biological models for change, as they indicate that the social circumstances of learning impact not only upon initial uptake of an invention but also on its eventual survival.

I suspect that Roux may not go far enough, and we need a much fuller recognition of the embedded nature of craft knowledge and the active contribution of the material environment to learning outcomes. I have found the developmental approach propounded by Wimsatt and Griesemer very useful in this regard, as they explicitly argue for the concept of scaffolding as an important (and largely missing) cog in any successful attempt at evolutionary explanation (Wimsatt and Griesemer 2007; Caporael et al. 2014). In so doing they draw on an important range of scholarship on scaffolding, from Andy Clark and Ed Hutchins to Lev Vygotsky and Donald Norman; these are many of the same sources of inspiration driving recent advances in cognitive archaeology (Knappett 2005; Malafouris 2013). The importance of acknowledging the constitutive role of the material environment lies in the implications for how we conceive of the learning of cultural traits. We might even have to admit that what from the outside may seem like a single cultural 'trait' is actually very variable, depending on how it is constituted in any given setting.

Even if I might have wished Roux to go further in the direction of extended cognition, I will nonetheless stick with her choice of area for technological study, as it seems well suited to exploring questions of knowledge and learning. The wheel technique requires the acquisition of complex skills over time, and there appear to be quite a lot of different responses to the suite of pros and cons it presents. Indeed, this analytical potential has been increasingly recognised by scholars working across many areas of the east Mediterranean, in large part inspired by the work of Roux in the Levant. Hence, we have seen the identification of the wheel-fashioning technique, as initially identified by Roux, on the island of Crete (Knappett 2004; Jeffra 2013), in the Cyclades (Gorogianni et al. 2016; Knappett forthcoming) and on the southern Greek mainland (Choleva 2012); and this technological perspective is also apparent in similar work in northern Greece, even if wheel fashioning *per se* has not yet been distinguished from wheel throwing (Kiriatzi 2000; Kiriatzi and Andreou in press). To these areas we can add Anatolia and Cyprus if we wish to create a fuller picture of exactly how and when the wheel technique appeared across the east Mediterranean (to the west of the Levant). The differences from one area to another are quite striking and provide a fascinating challenge to some of the assumptions of evolutionary approaches, much as anticipated by Roux (2013).

## 1 Cilicia, Crete and Euboea

We can begin with three quite separate regions that witness a fairly steady and widespread adoption of the wheel technique. The closest of these to the Levant is Cilicia, in southeast Turkey (see Fig. 1). Here the tournette is used from the Late Chalcolithic on, and so potters here were probably open to the technique of the potter's wheel when they began using it extensively in the later 3rd millennium BC (Türkteki 2013, 193). This adoption was probably through exposure to Levantine potters, who had been using the potter's wheel already for some time (Roux and Miroschedji 2009). Sites in Cilicia, such as Tarsus (Goldman 1956) and Kilise Tepe (Knappett and Kilikoglou 2007), see use of the wheel in Early Bronze (EB) III (see Fig. 1). The adoption here is probably a result of what Roux (2013) describes as 'cultural diffusion'—though one would have to examine more closely the social conditions behind the adoption of the technique in EB III and not earlier.

A few centuries later, in the early 2nd millennium BC, the wheel technique also appears on the island of Crete, quite a distance to the west (Fig. 1), and without any obvious signs of direct contact with the Levant or Cilicia. Though it is possible that some kind of contact with the Near East acted as a catalyst for the innovation (Knappett 1999), it otherwise seems to be an adoption driven largely by local factors. What is especially interesting about the wheel technique on Crete is the way in which potters all across the island employ it in very similar ways, gradually expanding its use until all pottery eventually, in the Late Bronze Age, is wheel fashioned. This too would then appear to be an example of Roux's 'cultural diffusion'. I provide detailed analysis of the Cretan case elsewhere (Knappett and van der Leeuw in press), and so I will not discuss further here.

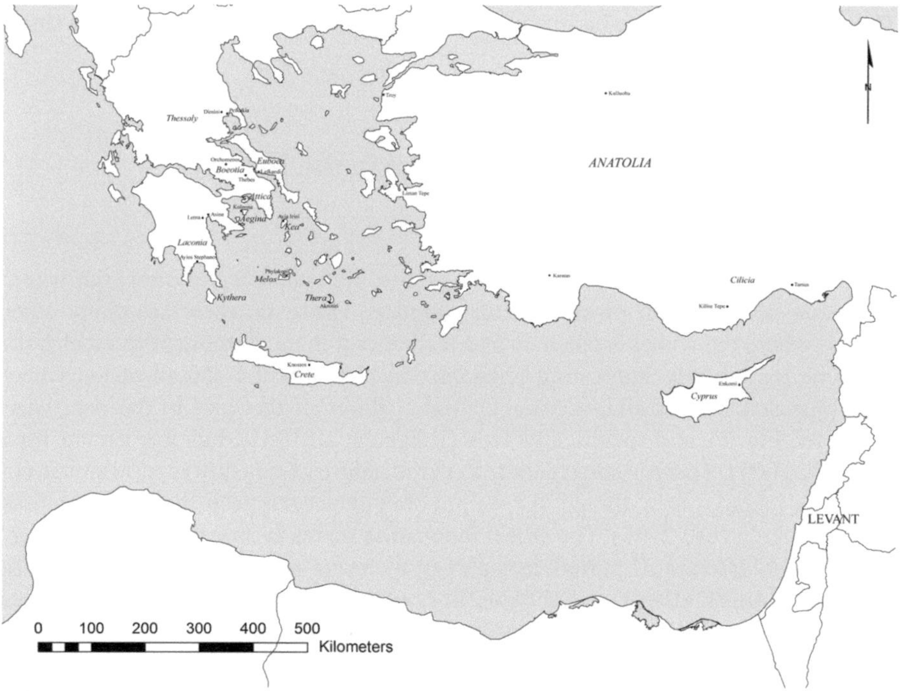


Fig. 1 Map of the east Mediterranean, showing sites mentioned in the text

The third example comes from the large island of Euboea, only separated from the southern Greek mainland by the narrowest of straits (Fig. 1). Many sites on the southern mainland only see a very partial adoption of the wheel, as we shall see below; but sites in Euboea and Boeotia, such as Lefkandi, Thebes and Orchomenos, and even further north in Thessaly, e.g. at Pefkakia and Dimini (Spencer 2010, 677), seem to take on the wheel technique more fully (Fig. 1). Moreover, this occurs relatively early, at more or less the same time as the adoption of the technique in Cilicia—and as we shall see, it may owe something to that region, indirectly. Yet, it seems quite distinct from, and much earlier than, the wheel innovation on Crete. Spencer's analysis of the pottery at Lefkandi (in Euboea) has allowed her to identify 8% of the Lefkandi I (late Early Bronze II) pottery as wheel made, which, though still a minority, is twice as much as at many other sites. What really stands out is the way the innovation then continues to develop at Lefkandi, with 20% wheel-made pottery in Early Helladic (EH) III and early Middle Helladic (MH) rising to 50% in MH I and II—when it is largely used for making a fine grey-burnished ware called 'Grey Minyan' (Spencer 2006). This pattern would seem to carry on into the later MH III and Late Helladic (LH) I periods too, across Boeotia and neighbouring areas (Mathioudaki 2011). Why does Lefkandi stand apart? Spencer attributes it to different modes of production, with more dispersed household production in the Peloponnese unable to sustain the wheel technique, unlike the more centralised workshops of central Greece (Spencer 2006). This does then beg the question of why production should be organised differently in central Greece in the first place. The Euboean case is quite hard to explain, especially when we see below the situation in other parts of the Aegean.

## 2 Western Anatolia and the Aegean in the Late 3rd Millennium BC

We have described above three regions where, for whatever reasons, the wheel technique does see full or almost full adoption. However, there are many other regions where the wheel is taken up as a technology in much more piecemeal fashion. One particularly interesting phenomenon concerns the spread of the wheel technique across Anatolia, seemingly from Cilicia all the way to the northwest coast (see Fig. 1), and specifically Troy (Blegen et al. 1951), but also Liman Tepe (Şahoğlu 2005). There is some debate as to the nature of this diffusion, whether via coastal or inland routes, though the weight of evidence suggests the inland route is far more likely (Efe 2007). The wheel innovation seems to spread at this time, during the Early Bronze III period, as a part of intensive connections between newly emergent political elites (see Şahoğlu 2014 on probable elite associations of depas cups, especially among travelling traders). What is especially interesting though is that the wheel technique sees a targeted rather than a widespread adoption: at the site of Küllüoba (Fig. 1), for example, in northwest Anatolia on the overland route from Cilicia to Troy, analysis by Murat Türkteki shows that only 3% of all the early

EB III pottery is wheel made, 8% in the following phase, and finally 13% in the latest of the Late EB III layers (Türkteki 2013, 194). The forms that are wheel made are very similar to those found at Troy (e.g. platter, tankard, depas), where wheel-made pottery was between 10 and 20%; at another notable site with EB III remains, Karataş (Fig. 1), wheel-made pottery is also rare (Türkteki 2013, 195). All of this points to an elite-driven adoption of the wheel, and the picture does not really line up very well with either of Roux's categories of cultural or demic diffusion.

This Anatolian use of the wheel has some impact further west, on the southern Greek mainland and neighbouring islands, at the end of EB II and into EB III. It coincides with a new assemblage consisting of five shapes—the *depas* cup, the tankard, the bell cup, the shallow bowl/plate and the beaked jug—all of which have antecedents in Anatolia. When appearing in the Aegean, this is called the Kastri/Lefkandi I group (see discussion of Lefkandi above). It is often assumed that these forms are wheel made, but this really only applies with any consistency to one type, the plate. Moreover, they only ever constitute a small percentage of the overall pottery assemblage at any given site, suggesting that a minority of potters had decided to use the wheel technique. At the site of Lerna in the Peloponnese (Fig. 1), for example, only 3% of the EH III pottery is wheel made; interestingly, it has also recently been established that the technique in use at Lerna was wheel fashioning (using coils) and not wheel throwing (Choleva 2012). A very similar situation has been observed at nearby Asine (Fig. 1), where wheel-made pots are a minority of 4–6% throughout the whole period from EH III to MH II (Spencer 2010). The numbers are so small that we should ask whether the wheel-made pots may be imports at these sites rather than locally produced (Zerner 1993; Spencer 2010). The answer to this seems to vary. For Lerna, Rutter (1995 464–66) suggests this category is, at least in part, made locally (although Spencer suggests that for the Peloponnese, there is little firm evidence to indicate any local use of the potter's wheel before MH II—Spencer 2006; Spencer 2010, 677). At Ayia Irini on the island of Kea, very close to the region of Attica on the southern mainland (Fig. 1), David Wilson was able to establish that at least some Kastri group types were locally produced, but that all the local versions were in fact handmade (Wilson 1999). The same seems to hold true at the site of Thebes in Boeotia (Fig. 1), where the Anatolianising wares are not only local, but also handmade (Hilditch et al. 2008).

## 3 The Aegean in the 2nd Millennium BC

When we turn to the beginning of the 2nd millennium BC, there is a real shift in the distribution of wheel technologies across the Aegean. As already noted above, this is because c. 1900 BC on Crete potters starts to use the wheel in such a way that it quite quickly and widely catches on (Knappett 1999; Knappett and van der Leeuw in press). However, this has very little connection with the Early Bronze Age appearance of the wheel in areas further north, which had no discernible impact on Crete. Instead, what we see on Crete is largely a local development, perhaps with some inspiration from Near Eastern connections (Knappett 1999). At the same time, however, a quite different tradition of wheel use continues in central Greece in Euboea and Boeotia, evolving from the Early Helladic III innovation, and entirely unconnected to Crete. This continuity is the exception rather than the rule, however, so that in the early Middle Bronze Age, we have a wide zone between these northern (Euboea/Boeotia) and southern (Crete) areas where the wheel is generally overlooked, ignored or resisted. What is then quite interesting to observe is how, where and why the wheel is then adopted (and in some cases, like Kea, for a second time) in some of these areas. In some places, we actually end up with two different kinds of wheel-made pottery side by side—from both the Euboean/Boeotian tradition (i.e. Grey Minyan ware) and the Cretan (i.e. Lustrous Decorated ware). Indeed, at the site of Ayios Stephanos, in Laconia (Fig. 1), some types of dark Minyan ware, evidently a central Greek form, are made using the wheel technique (Spencer 2006, 193). However, the site is also a consumer of Lustrous Decorated ware, which initially is of Cretan inspiration, though this wheel-made Lustrous Decorated pottery seems to come in large part from Kythera (Kiriatzi 2010), an island very much in the Cretan orbit, and probably colonised from there in the Early Bronze Age (Broodbank and Kiriatzi 2007). There is debate as to whether some of it might also be locally produced in Laconia (see Dickinson 2014), but one way or another potters at Ayios Stephanos begin to use the wheel in a selective fashion during MH II and quite possibly inspired from the south rather than the north.[3]

In the Cyclades, however, there is not really any question that when we see the uptake of the wheel, it is clearly through contact with the south and not the north. At Kea, for example, where wheel-made pottery is used (if not made, see above) in the Early Bronze Age, the reappearance of wheel-made pottery, this time definitely locally produced, in Period IV (i.e. the middle of the Middle Bronze Age) seems to owe everything to Crete and nothing to Euboea/Lefkandi (Gorogianni et al. 2016). These same authors mention that a similar situation is observed on the nearby island of Aegina (Fig. 1), where there is also a rare use of the potter's wheel. Indeed, its occurrence in Phase I at the site of Kolonna on Aegina is associated only with the very few locally made Minoan-style forms, though wheel-made Grey Minyan imports also crop up (Gauss and Smetana 2007, 63; Gauss and Kiriatzi 2011, 251–2). It is also interesting that Aegina saw no uptake of the wheel technique in EB II–III, despite its centrality in regional exchange networks between the southern Greek mainland and the Cyclades. As for the later use of the wheel on Aegina, it seems to again emerge in connection with the local manufacture of Mycenaean-style pottery in LB I, though with little influence on local-style wares until later, in LB III.[4]

---

[3] Work in progress by Maria Choleva should throw light on the question of the emergence of the wheel-fashioning technique in the Peloponnese during the Middle Helladic period.

[4] With many thanks to Evangelia Kiriatzi and Walter Gauss, Personal communication in 2015

When we turn from Kea and Aegina to other islands further south, notably the Cycladic islands of Melos and Thera (Fig. 1), we see yet another scenario. In the middle of the Middle Bronze Age, equivalent to Period IV on Kea, there is no use of the wheel technique on either island (Berg 2007; Knappett and Nikolakopoulou 2005, 2008). Cretan ceramic imports certainly occur in some quantities at the site of Akrotiri on Thera and to a lesser extent at Phylakopi on Melos too. So there was definitely contact between these regions—and it is quite likely that potters on Thera and Melos had every opportunity to learn about the wheel technique from Crete. Yet, there seems to be little interest in its adoption. However, the potter's wheel *is* introduced in the *later* MBA at both sites (equivalent to Middle Minoan IIIA on Crete, c. 1700 BC), at the same time that it begins to expand in use on Kea too (in Period V at Ayia Irini, Kea). Now the number of imports from Crete rises significantly, and there is a simultaneous move towards limited local imitation of Cretan styles. The wheel technique is restricted to the manufacture of a handful of recognisably Cretan types, namely, plain ledge-rim bowls and straight-sided cups. When set against the total resistance to the wheel in earlier phases, this sudden uptake of the technique on Melos and Thera is quite telling. It suggests that interactions between Crete and the Cyclades had been transformed. It may be that the uptake of the technique was elite driven, and this would certainly fit with the rapid and only partial nature of its adoption. Nonetheless, the technique sees fuller adoption than in other seemingly elite-driven contexts, as the wheel is gradually used in the production of more and more forms into the early Late Bronze Age period. This could suggest not only that there was mobility of craftspeople enabling the technological transfer but also that the wheel technique did become quite well embedded in local communities of practice (Abell and Hilditch 2016). Still, the wheel technique does not bed in quite as fully as it did on Crete, with much of the local pottery repertoire continuing to be made using coil-building techniques, with little use of any kind of rotation even for finishing. This is perhaps in part due to the shift in influence during LB I from Crete to the mainland. We cannot track this shift at all at Akrotiri, because it is exactly the time when the site is engulfed by the Theran eruption in the Late Minoan IA phase (the absolute chronology is hotly debated, but this is either in the late 17th or late 16th c. BC) and not reoccupied for centuries. However, Melos and Ayia Irini do both see continued occupation into Late Cycladic II–III, so we can say something about technological choices. Abell and Hilditch (2016) observe that, at least on Kea, even though in LC II–III there is a lot of interest in the consumption of Mycenaean fine pottery, local pottery production seems largely unconnected with Mycenaean practices. This marks quite a change from the preceding 'Minoanising' periods when production choices did seem quite geared to Minoan habits. If the continued development of the wheel technique in the Cyclades would have relied upon stable links between communities of practice on Crete and the Cyclades, then one can imagine how the switch of political influence to the mainland might have curtailed this technological trajectory. However, we could certainly learn more about the wheel in Late Cycladic III—as Abell and Hilditch note (2016), there has been a lot less research on forming techniques in these later periods.

## 4 Cyprus

The examples cited so far put the lie to the idea that the spread of the potter's wheel was a regular and predictable diffusion 'by contagion' from east to west. Nowhere is this truer than the island of Cyprus (Fig. 1), which is completely bypassed in this supposed spread both at the end of the 3rd millennium BC and in the early 2nd millennium BC. Despite its proximity to the Levant, and indeed to Cilicia, Cypriot pottery production remains resolutely handmade. It is not until centuries later, c. 1650 BC, that the wheel makes an appearance, with a rather sporadic uptake, connected with a period of increased interregional contacts, though with Egypt and the Near East rather than Anatolia (Crewe and Knappett 2012; Crewe 2007a; Crewe 2007b). Moreover, its adoption is linked, it would seem, with emulation of foreign consumption practices in the Late Cypriot I period. New coastal settlements are established, seemingly to take advantage of opportunities for exchange. One such site is Enkomi (Fig. 1), and this and other coastal 'emporia' testify to new social forms and increasing wealth (Crewe 2007a). While wheel fashioning is an innovation that pops up at a number of these new sites, there seems to be little connection between them and a strong continuing tradition of handmade wares, such as White Slip and Base Ring wares, for example. This scenario is one with little sharing of technological knowledge across the island, coupled with the persistence of strong regional identities; this contrasts sharply with what is observed on Crete (Crewe and Knappett 2012). Indeed in Late Cypriot II, handmade pottery styles such as White Slip and Base Ring increase in frequency, used both locally and exported off-island. This fragility in the long-term resilience of the technique would seem compatible with a scenario of elite-driven innovation.

## 5 Discussion: Innovation 'Failure'?

If we were to begin with the assumption that the wheel technique offers obvious advantages over hand-building techniques in terms of both the quality and quantity of output, then we might expect to see a regular, smooth uptake of the innovation as soon as it becomes available to ancient potters. That is, if they are in contact with potters who have the wheel, then they will surely copy that technique. Yet, when we look at the evidence from across the east Mediterranean in the Early and Middle Bronze Age, what we see is anything but a predictable adoption of the technology of the potter's wheel. The start-stop pattern of partial adoption in many regions—especially in northwest Anatolia, the Cyclades, Cyprus and parts of the southern Greek mainland—indicates that 'macroevolutionary' expectations are confounded and that instead we have to turn to microevolutionary processes for explanation and notably the obvious sensitivity of the innovation to local context. What this really means is that the technique does not have the exact same status irrespective of location, as might a package of information that can be repeatedly uploaded and downloaded into different storage devices. The technique, it would seem, takes on a different status depending on precisely *how* it is learnt—and we must stress that it is a technique that is difficult to acquire, usually requiring some kind of apprenticeship (see Roux and Corbetta 1990; also Wendrich 2012 on apprenticeship more broadly). And given this difficulty, it seems likely that there would exist different strategies for learning, which we might understand in terms of various mixes of declarative and nondeclarative knowledge (see Knappett and van der Leeuw in press). Ideally, a long-term apprenticeship would embed craft knowledge deeply through nondeclarative learning—and this would presumably make that knowledge very resilient over time, assuming it was also passed on in that way to subsequent generations. One might also suppose such learning to lean heavily on the environment, through scaffolding structures both social and material in character (see Caporael et al. 2014). This is not to say that such learning does not also involve some degree of declarative knowledge. By the same token, we could well imagine that learning predicated largely on declarative knowledge might shorten the time of learning, but might also make that knowledge acquired somewhat more fragile—also in part, surely, connected to a much lesser investment in the scaffolding of the socio-material environment to distribute the cognitive load of learning.

I argue elsewhere (Knappett and van der Leeuw in press) that we might hypothesise that elite-driven innovation would lean more towards declarative rather than nondeclarative learning; and hence, one could expect such innovation to be quite quick, yet also more fragile over time, as well as, presumably, affecting fewer potters, if the elites were only interested in production of particular wares for their own consumption. Picking up a technique through declarative knowledge spread by elites may not then be a sustainable strategy for widespread adoption (as well as being less scaffolded and less embedded)—though this may very well also be exactly what an elite had in mind. Therefore, when we see the kind of partial and ultimately fragile uptake documented here in many regions, we might tentatively attribute it to this dynamic of elite-driven, non-embedded, declarative learning. This then suggests that, by contrast, the process in Crete, Cilicia and Euboea was somewhat different in being more embedded and scaffolded and less contingent upon elite demands.

## 6 Conclusions

If we are then arguing that a variety of strategies may exist for learning a 'single' technique such as the potter's wheel, and that these strategies may rely differentially on knowledge embedded in the environment (i.e. scaffolding), then what are the implications for evolutionary approaches to cultural transmission? It should at the very least make us question the notion of the potter's wheel as a 'cultural trait', viewed as part of a design grammar, a module that can be inserted or removed. We should also be moved to question the assumption that the learning of wheel technology involves simply the transfer of information stored in human brains. Much of the work on apprenticeship does after all show how the socio-material environment itself is used in learning, such that craft knowledge cannot easily be reduced to information. It is not the case that the exact paths of information transfer are irrelevant, as long as the information is transmitted. With a complex technology that requires learning, the strategy employed for learning does appear to have long-term consequences.

In what other ways could it be said that evolutionary models do not sit well with the diffusion of the wheel technique across the east Mediterranean? Well, we might also raise the favoured metaphor in biological models of the branching tree. There are so many localised patterns in adoption and rejection across the Mediterranean that we would have great difficulty in making the pattern of spread look much like a tree—we could more readily imagine a network rather than a tree. This is not necessarily a fatal blow to Darwinian approaches, as the recognition of lateral gene transfer has already necessitated the adoption of network models in addition to tree models in biology (Dagan 2011). Moreover, lateral gene transfer has a corollary in cultural evolution that has been dubbed 'combinatorial evolution', very aptly employed to explain sudden, non-branching episodes of technological change (cf. Arthur 2009).

This paper has been somewhat critical of evolutionary approaches, and the reader may wonder why it engages with them at all. What is to be gained from just saying that an example like the potter's wheel confounds most of the expectations of typical biological models as applied to culture—especially when no alternative is provided? My aim has been to recognise that evolutionary approaches do us a great service by asking those broader questions that interpretive archaeologies often fail to. As Wimsatt and Griesemer (2007, 227) put it, critiquing 'thick' views of culture (the equivalent to interpretive here): 'most of these writers reject any attempt to give an evolutionary account of the origins of culture or the nature of culture change'. On the other hand, they are equally critical of those biological approaches that are too 'thin', failing to take learning seriously by 'black boxing' human ontogeny (Wimsatt and Griesemer 2007, 227). They propose instead an approach that is neither thick nor thin, but of 'medium viscosity'. This paper follows in this very spirit. Surely we do need reductive approaches (because we learn from comparing cultural traditions across time and space), while also doing justice to the complexities in the data. Our focus here on craft learning and technological traditions serves to reinforce the message offered by Wimsatt and colleagues, with its strong orientation towards cognitive development and the recognition of scaffolding as a key process therein that evolutionary approaches limit themselves when they imagine that 'human technological traditions…consist of information stored in human brains' (cf. Jordan 2014).

**Acknowledgements** I am grateful to Larissa Mendoza Straffon for her patience in waiting for my final draft. Thanks also to Iro Mathioudaki, Maria Choleva, Vasif Şahoğlu and Evangelia Kiriatzi for their invaluable advice on pottery technologies in mainland Greece and Anatolia.

## References

Abell, N., & Hilditch, J. (2016). Adoption and adaptation in pottery production practices: investigating Cycladic community interactions through the ceramic record of the second millennium BC. In P. Pavuk, E. Gorogianni, & L. Girella (Eds.), *Minoanisation and Mycenaeanisation.* Oxford: Oxbow Books.

Arthur, W. B. (2009). *The nature of technology: What it is and how it evolves.* New York: Free Press.

Berg, I. (2007). Meaning in the making: The potter's wheel at Phylakopi, Melos (Greece). *Journal of Anthropological Archaeology, 26*(2), 234–252.

Blegen, C. W., Caskey, J. L., & Rawson, M. (1951). *Troy, the third, fourth and fifth settlements.* Princeton, NJ: Princeton University Press.

Broodbank, C., & Kiriatzi, E. (2007). The first 'Minoans' of Kythera re-visited: Technology, demography and landscape in the Early Bronze Age Aegean. *American Journal of Archaeology, 111*(2), 241–274.

Caporael, L. R., Griesemer, J. R., & Wimsatt, W. C. (2014). Developing scaffolds: An introduction. In L. R. Caporael, J. R. Griesemer, & W. C. Wimsatt (Eds.), *Developing scaffolds in evolution, culture, and cognition* (pp. 1–20). Cambridge: MIT Press.

Choleva, M. (2012). The first wheelmade pottery at Lerna: Wheel-thrown or wheel-fashioned? *Hesperia, 81*, 343–381.

Crewe, L. (2007a). *Early Enkomi. Regionalism, trade and society at the beginning of the Late Bronze Age on Cyprus.* Oxford: BAR IS 1706.

Crewe, L. (2007b). Sophistication in simplicity: The first production of wheel-made pottery on Late Bronze Age Cyprus. *Journal of Mediterranean Archaeology, 20*(2), 209–238.

Crewe, L., & Knappett, C. (2012). Technological innovation and island societies: wheel-made pottery on Bronze and Iron Age Crete and Cyprus. In G. Cadogan, M. Iacovou, K. Kopaka, & J. Whitley (Eds.), *Parallel lives. Ancient island societies in Crete and Cyprus. Proceedings of the conference held at Nicosia 1-3 December 2006* (pp. 175–185). London: BSA Supplementary Volume.

Dagan, T. (2011). Phylogenomic networks. *Trends in Microbiology, 19*(10), 483–91.

Dickinson, O. T. P. (2014). Late Helladic I revisited: the Kytheran connection. In D. Nakassis, J. Gulizio & S. A. James (Eds.), *KE-RA-ME-JA: Studies presented to Cynthia W. Shelmerdine* (pp. 3–15). Philadelphia: INSTAP Academic Press.

Efe, T. (2007). The theories of the 'Great Caravan Route' between Cilicia and Troy: The Early Bronze Age III period in inland western Anatolia. *Anatolian Studies, 57*, 47–64.

Gardner, A., & Cochrane, E. (2011). Evolutionary and interpretive archaeologies: A discussion. In E. Cochrane & A. Gardner (Eds.), *Evolutionary and interpretive archaeologies: A dialogue* (pp. 11–29). Walnut Creek: Left Coast Press.

Gauss, W., & Kiriatzi, E. (2011). *Pottery production and supply at Bronze Age Kolonna, Aegina: an integrated archaeological and scientific study of a ceramic landscape.* Ägina Kolonna, Forschungen und Ergebnisse 5. Contributions to the Chronology of the Eastern Mediterranean. Vienna: Österreichische Akademie der Wissenschaften.

Gauss, W., & Smetana, R. (2007). Aegina Kolonna, the ceramic sequence of the SCIEM 2000 project. In F. Felten, W. Gauss & R. Smetana (Eds.), *Middle Helladic pottery and synchronisms* (pp. 57–80). Vienna: Austrian Academy of Sciences.

Goldman, H. (1956). *Excavations at Gözlükule, Tarsus, Vol 2: the Neolithic through the Bronze Age.* Princeton.

Gorogianni, E., Abell, N., & Hilditch, J. (2016). Reconsidering technological transmission: the introduction of the potter's wheel at Ayia Irini, Kea, Greece. *American Journal of Archaeology.*120(2)

Hilditch, J., Kiriatzi, E., Psaraki, K., & Aravantinos, V. (2008). Early Helladic II pottery from Thebes: An integrated typological, technological and provenance study. In Y. Facorellis, N. Zacharias, K. Polikreti, T. Vakoulis, Y. Bassiakos, V. Kiriatzi, & E. Aloupi (Eds.) *Archaeometry studies in the aegean: Reviews and recent developments. Proceedings of the 4th*

*International Symposium of the Hellenic Society for Archaeometry* (pp. 263–268). Oxford: Archaeopress.

Jeffra, C. (2013). Are-examination of early wheel potting in Crete. *Annual of the British School at Athens, 108*, 31–49.

Jordan, P. (2014). *Technology as human social tradition: Cultural transmission among hunter-gatherers.* Oakland: University of California Press.

Kiriatzi, E. (2000). *Ceramic technology and production. The pottery of Late Bronze Age from Toumba Thessaloniki.* PhD thesis. Aristotle University of Thessaloniki. (in Greek).

Kiriatzi, E. (2010). Minoanising pottery traditions in southwest Aegean during the Middle Bronze Age: Understanding the social context of technological and consumption practice. In A. Philippa-Touchais, G. Touchais, S. Voutsaki & J. Wright (Eds.), *Mesohelladika. La Grèce continentale au Bronze moyen* (pp. 683–699). Athens: BCH Suppl. 52.

Kiriatzi, E., & Andreou, S. (In press). Mycenaean and Mycenaeanising pottery across the Mediterranean: A multi-scalar approach to technological mobility, transmission and appropriation. In E. Kiriatzi % C. Knappett (Eds.), *Human mobility and technological transfer in the prehistoric Mediterranean.* British School at Athens Studies in Greek Antiquity 1. Cambridge: CUP.

Knappett, C. (1999). Tradition and innovation in pottery forming technology: Wheel-throwing at Middle Minoan Knossos. *Annual of the British School at Athens, 94*, 101–129.

Knappett, C. (2004). Technological innovation and social diversity at Middle Minoan Knossos. In G. Cadogan, E. Hatzaki & A. Vasilakis (Eds.), *Knossos: palace, city, state* (pp. 257–265). London: BSA Studies 12.

Knappett, C. (2005). *Thinking through material culture: An interdisciplinary perspective.* Philadelphia: University of Pennsylvania Press. Liège and Austin

Knappett, C. (forthcoming). The imported Minoan pottery, In I. Nikolakopoulou (Ed.), *The Middle Cycladic pottery from Akrotiri, Thera.* Athens: Archaeological Society of Athens.

Knappett, C., & Kilikoglou, V. (2007). Pottery fabrics and technology. In J. N. Postgate & D. Thomas (Eds.), *Excavations at Kilise Tepe 1994-98: From Bronze Age to Byzantine in Western Cilicia* (pp. 241–72). Cambridge: McDonald Institute Monographs.

Knappett, C., & Nikolakopoulou, I. (2005). MBA interaction networks: Crete, the Cyclades and the South-East Aegean. In R. Laffineur & E. Greco (eds.), *Emporia. Aegeans in central and eastern Mediterranean, Proceedings of the 10th international Aegean conference, Athens, Italian School of Archaeology, 14–18 April 2004* (pp. 175–183). Aegaeum 25. Liège and Austin.

Knappett, C., & Nikolakopoulou, I. (2008). Colonialism without colonies? A Bronze Age case study from Akrotiri, Thera. *Hesperia, 77*, 1–42.

Knappett, C., & van der Leeuw, S. E. (In press). A developmental approach to ancient innovation: The potter's wheel in the Bronze Age East Mediterranean. Special issue of *Pragmatics and Cognition.*

Malafouris, L. (2013). *How things shape the mind: A theory of material engagement.* Cambridge: MIT Press.

Mathioudaki, I. (2011). *Mainland polychrome pottery in mainland Greece and the Aegean.* Unpublished PhD Thesis, University of Athens.

Roux, V. (2010). Technological innovations and developmental trajectories: social factors as evolutionary forces. In M. J. O'Brien & S. J. Shennan (Eds.), *Innovation in cultural systems. Contributions from evolutionary anthropology* (pp. 217–234). Vienna Series in Theoretical Biology. Cambridge: MIT Press.

Roux, V. (2013). Spreading of innovative technical traits and cumulative technical evolution: Continuity or discontinuity. *Journal of Archaeological Method and Theory, 20*, 312–330.

Roux, V., & Bril, B. (Eds.) (2005). *Stone knapping: The necessary conditions for a uniquely hominin behaviour.* Cambridge: McDonald Institute for Archaeological Research.

Roux, V., & Corbetta, D. (1990). *The potter's wheel. Craft specialisation and technical competence.* New Delhi: Oxford and IBH Publishing.

Roux, V., & de Miroschedji, P. (2009). Revisiting the history of the potter's wheel in the southern Levant. *Levant, 41*, 155–173.

Rutter, J. (1995). *Lerna. A Preclassical Site in the Argolid, III: The Pottery of Lerna IV.* Princeton: Princeton University Press.

Şahoğlu, V. (2005). The *Anatolian trade network* and the Izmir region during the Early Bronze Age. *Oxford Journal of Archaeology, 24*, 339–361.

Şahoğlu, V. (2014). The *depas* and tankard vessels. In M. Lebeau (Ed.), *ARCANE Interregional, Vol. I. Ceramics* (pp. 289–311). Turnhout: Brepols.

Spencer, L. (2006). *Pottery technology and socio-economic diversity on the Early Helladic III to Middle Helladic II Greek mainland.* Unpublished PhD thesis, Institute of Archaeology, University College London.

Spencer, L. (2010). The regional specialization of ceramic production in the EH III through MH II period. In A. Philippa-Touchais, G. Touchais, S. Voutsaki & J. Wright (Eds.), *Mesohelladika. La Grèce continentale au Bronze moyen* (pp. 669–81). Athens: BCH Suppl. 52.

Türkteki, M. (2013). The first use of wheel-made pottery and its distribution in western and central Anatolia. In L. Bombardieri, A. D'Agostino, G. Guarducci, V. Orsi & S. Valentini (Eds.), *SOMA 2012. Identity and Connectivity: Proceedings of the 16th Symposium on Mediterranean Archaeology, Florence, Italy, 1-3 March 2012, Volume 1* (pp. 193–196). Oxford: BAR-IS 2581(I).

Wendrich, W. (2012). Archaeology and apprenticeship: Body knowledge, identity, and communities of practice. In W. Wendrich (Ed.), *Archaeology and apprenticeship: Body knowledge, identity and communities of practice* (pp. 1–19). Tucson: The University of Arizona Press.

Wengrow, D. (2011). Cognition, materiality and monsters: The cultural transmission of counter-intuitive forms in Bronze Age societies. *Journal of Material Culture, 16*(2), 131–149.

Wilson, D. E. (1999). *Ayia Irini: Periods I-III, The Early Neolithic and Early Bronze Age Settlements. Part 1. Pottery and Small Finds.* Mainz: Phillip von Zabern.

Wimsatt, W. C., & Griesemer, J. (2007). Reproducing entrenchments to scaffold culture: the central role of development in cultural evolution. In R. Sansom & R. Brandon (Eds.), *Integrating evolution and development: From theory to practice* (pp. 227–323). Cambridge: MIT Press.

Zerner, C. (1993). New perspectives on trade in the Middle and early Late Helladic periods on the mainland. In C. Zerner, P. Zerner & J. Winder (Eds.), *Wace and blegen: Pottery as evidence for trade in the aegean bronze age, 1939-1989* (pp. 39–56). Amsterdam: J.C. Gieben.

# Mosaic Evolution in Cultural Frameworks: Skateboard Decks and Projectile Points

**Anna Marie Prentiss, Matthew J. Walsh, Randall R. Skelton, and Matt Mattes**

**Abstract** There has been significant debate in paleoanthropology and more recently, archaeology, over the concept of mosaic evolution. Essentially, proponents of the concept argue that different aspects of organisms evolve separately while others argue that organisms evolve as integrated entities. Similarly, archaeologists debate the relevance of cultural evolution as a complex multi-scalar process. In this paper we conduct two cladistic analyses of cultural phenomena focusing on skateboard decks and projectile points from an archaeological site to examine variability in the evolutionary process. We find evidence for mosaic evolution in both studies and conclude that modularity likely is an important factor in cultural evolution, at least at the level of artifact design. We caution future investigators of evolution in ancient stone tools that modularity could have complicating effects on phylogenetic outcomes unless explicitly considered.

**Keywords** phylogenetic analysis • mosaic evolution • projectile points

## 1 Introduction

Evolutionary studies of artifacts have progressed in a number of significant ways in recent years. We recognize that artifact evolution can be understood in strict neo-Darwinian terms as a process by which variation arising from innovation and copying errors is sorted in the long term by selection and drift (Goodale et al. 2011) giving rise to phylogenetic trees characterized by high rates of branching (Jordan and Shennan 2009; Prentiss et al. 2011; Tehrani 2011; Tehrani and Collard 2002). But we have also learned that artifact evolution inevitably is affected by a range of tokogenetic processes and the effects of the Hannah Principle (Eldredge 2006). Tokogenetic signals result from lateral transfer of characters as might occur in cultural borrowing of ideas or horizontal cultural transmission. The Hannah Principle recognizes the

A.M. Prentiss (✉) • M.J. Walsh • R.R. Skelton • M. Mattes
Department of Anthropology, The University of Montana, Missoula, MT 59812, USA
e-mail: anna.prentiss@umontana.edu

 
L. Mendoza Straffon (ed.), *Cultural Phylogenetics: Concepts and Applications in Archaeology*, Interdisciplinary Evolution Research,
DOI 10.1007/978-3-319-25928-4_6

effects of different groups developing independent solutions to common problems. Thus, phylogenies may contain reticulations and cladograms are compounded by frequent homoplasies. Further, technologies can behave over time in ways quite different from biological species. For example, extinction is not necessarily always final. Some items are resurrected, while others go dormant, only to be resurrected when wider conditions change (Chatters 2009; Eldredge 2009). In this paper, we wish to explore another facet of evolutionary process, common to biological and cultural systems, known as mosaic evolution (Gould 1977).

Mosaic evolution is today widely recognized in the evolutionary literature and essentially refers to the process of independent change in different portions of a phenotype (Stanley 1998). Mosaic evolution is well known from the phylogenies of a number of genera. Probably the most widely cited example comes from early hominin evolution where bipedal locomotion evolved rapidly during a period in which evolution of the cranium was minimal (McHenry 1994). Other examples from paleoanthropology include dental reduction, postcranial gracilization, and encephalization (Brace 1995; Skelton and McHenry 1998). Research in human genetics is also confirming mosaic evolution (Pääbo 2003). Mosaic evolution is, of course, not confined to humans as it has been documented in the evolutionary histories of many extant species including elephants (Maglio 1973) and birds (Clarke and Middleton 2008) and extinct genera such as pterosaurs (Lü et al. 2010). Consequently, despite some recent debates (e.g., Kemp 2007a, b), understanding the processes of mosaic or modular evolution is an issue of ongoing importance in evolutionary research (Brandon 1999; Schlosser 2002).

Scholars of material cultural evolution have known for some time that artifacts can evolve in a modular fashion. Boyd et al. (1997) effectively predicted it when they proposed their model of "cultures as assemblages of many coherent units." Eldredge (2000, 2011) demonstrates mosaic evolution in cornets especially associated with changes in bells and valves. Dagg (2011) finds similar evidence for mosaic evolution in mousetraps, suggesting that the ability to recombine parts represents an essential requirement for culture to be cumulative (e.g., Richerson and Boyd 2005). Evolutionary archaeologists have implicitly recognized mosaic evolution, for example, in projectile points, noting that haft modifications often evolve independent of blade configurations (Darwent and O'Brien 2006; O'Brien et al. 2001). Despite this knowledge, there has not been any significant research into effects of mosaic evolution on the construction of cultural phylogenies. If we do not fully understand the effects of mosaic evolution, then we may be making errors in some phylogenetic reconstructions.

In this study we test for the effects of mosaic evolution focusing on a well-understood item, the skateboard deck (Prentiss et al. 2011). We apply a cladogenetic approach drawing in particular on the work of Skelton and McHenry (1998) and then seek to develop implications for future studies of more ancient technologies that are far less well known. As an example, we follow with an archaeological case study of projectile point morphology designed to look for effects of mosaic evolution. Projectile points have been frequently studied by archaeologists interested in phylogenetic histories (Buchanan and Collard 2007; Darwent and O'Brien 2006; O'Brien et al. 2001; O'Brien and Lyman 2003).

## 2 Problem, Materials, and Methods

### 2.1 *Skateboard Decks*

In a previous phylogenetic study, Prentiss et al. (2011) constructed phylogenetic trees that effectively replicated the evolution of the skateboard deck as it is known from other historical sources (Goodrich 2010; Weyland 2002). Briefly, the first professional skateboard deck, known as the Makaha Phil Edwards, developed in 1963, was little more than a small flat piece of wood loosely shaped like a surfboard. This design was highly influential and effectively persists to this day. However, a wide range of variation in more specialized designs developed during the mid-1970s that included slalom racing boards with narrow ends, a wide midsection, and cambered form; bowl riders with wide shapes and large rear kicktails; freestyle boards, much closer to the Makaha Phil Edwards but with a kicktail and longer nose area; and long downhill boards, originally cut from water skis, coming in a variety of shapes and all marked by length of typically greater than 36 in. The explosion in designs in the 1970s has been likened by skateboard historians (and nonbiologists) such as James Weyland (2002), as analogous to the Cambrian explosion. An additional form, developed in the early 1990s known as the "popsicle stick board," was designed for flexible shifting between street and wall riding and featured the now widely recognized popsicle stick shape with double kicktails and convex deck surface. Cladistic and network analysis of 17 boards by Prentiss et al. (2011) confirmed a branching pattern reflecting the emergence of slalom, bowl rider, freestyle, and downhill and street board designs (Figs. 1 and 2). Embedded within the branching pattern was a strong pattern of borrowing of ideas. Critically, the street boards were clearly the most intensely reticulated reflecting their actual history as blended designs (Fig. 2).

As is typical of other artifact evolution studies, Prentiss et al. (2011) did not seek to assess potential impacts of mosaic evolution on the development of their skateboard deck phylogenies. However, there is good reason to believe that differential evolution of particular characters likely occurred and that this could in turn have impacted the structure of the phylogenies. Character states associated with end shape, width, kicktails, and cambering likely developed to solve particular problems and likely evolved as modules. In order to test for the effects of this phenomenon, we reanalyzed a smaller but approximately equivalent data base derived from Prentiss et al. (2011) and compared the results to trees derived from sets of traits best describing four functionally distinct board designs (Table 1). This is effectively the approach outlined by Skelton and McHenry (1998), who examined bias from differential measurement of heavy chewing and other characters on the development of hominid phylogenies. We relied upon PAST 2.04 (Hammer et al. 2001) to derive trees and associated descriptive statistics (consistency and retention indices; Table 2) using strict consensus parsimony methods. All cladograms were rooted with the Makaha Phil Edwards board.

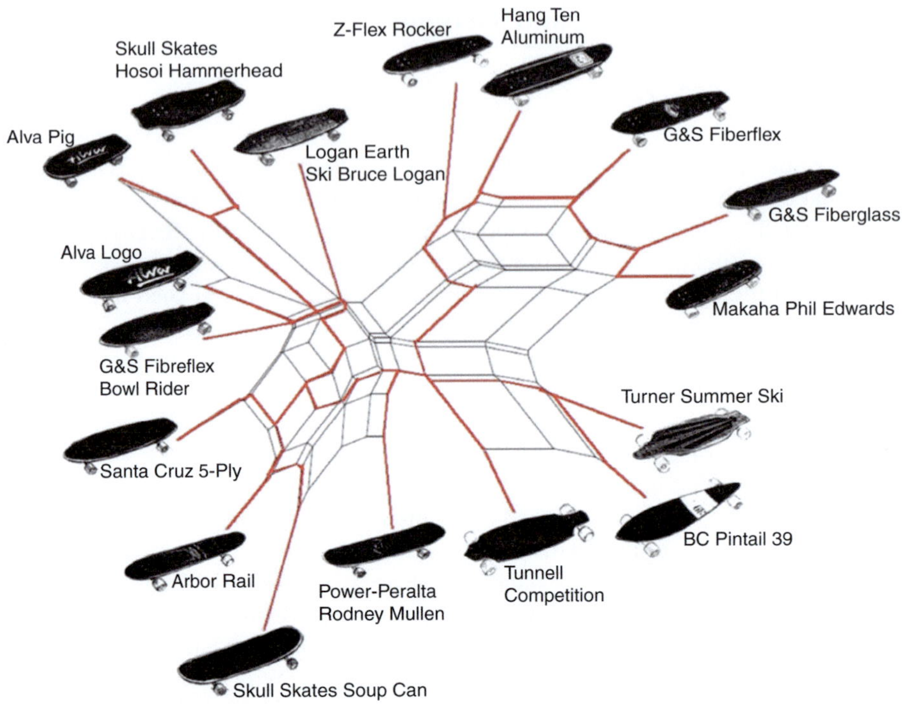


**Fig. 1** Splitstree network illustrating branching and blending in skateboard evolution (From Prentiss et al. (2011))

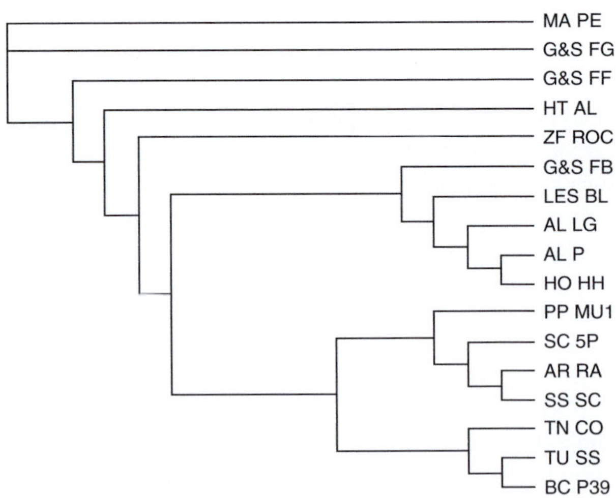


**Fig. 2** PAUP neighbor-joining tree (From Prentiss et al. (2011))

**Table 1** Data used to derive skateboard study cladograms

| Total data (length 28–34, length >34, width 7–9, width >9, nose 4+, T 4+, kicktail standard, kicktail front, concave, camber, maximum width front-center, minimum width nose, wheel wells) | | |
|---|---|---|
| Maker | | Model |
| MA | PE | 0000000000010 |
| LES | BL | 1010011000010 |
| AR | RA | 0110011000100 |
| TN | CO | 0110010000010 |
| PP | MU | 0010111000010 |
| SS | SC | 1010111110000 |
| AL | P | 1001011000111 |
| AL | LG | 1010011000111 |
| Slalom function data (length 28–34, length >34, width 7–9, nose 4+, T 4+, camber, maximum width front-center, minimum width nose) | | |
| Maker | | Model |
| MA | PE | 00000001 |
| LES | BL | 01101011 |
| AR | RA | 10111010 |
| TN | CO | 10100100 |
| PP | MU | 00111011 |
| SS | SC | 01111010 |
| AL | P | 01101011 |
| AL | LG | 01101011 |
| Bowl rider function data (length 28–34, width >9, nose 4+, T 4+, kicktail standard, concave, maximum width front-center, minimum width nose, wheel wells) | | |
| Maker | | Model |
| MA | PE | 000000010 |
| LES | BL | 100110110 |
| AR | RA | 001110100 |
| TN | CO | 000000000 |
| PP | MU | 001110110 |
| SS | SC | 101111100 |
| AL | P | 110110111 |
| AL | LG | 100110111 |
| Freestyle function data (length 28–34, width 7–9, nose 4+, T 4+, kicktail standard, concave, camber, maximum width front-center, minimum width nose, wheel wells) | | |
| Maker | | Model |
| MA | PE | 00000010 |
| LES | BL | 11111110 |
| AR | RA | 11111100 |
| TN | CO | 11000000 |
| PP | MU | 01111110 |
| SS | SC | 11111000 |
| AL | P | 10011111 |
| AL | LG | 11011111 |

(continued)

**Table 1** (continued)

| Street function data (length >34, width 7–9, nose 4+, T 4+, kicktail standard, concave, camber, maximum width front-center, minimum width nose) | | |
|---|---|---|
| Maker | | Model |
| MA | PE | 000000001 |
| LES | BL | 011110011 |
| AR | RA | 111110010 |
| TN | CO | 110000000 |
| PP | MU | 011110011 |
| SS | SC | 011111100 |
| AL | P | 000110011 |
| AL | LG | 010110011 |

*MAPE* Makaha Phil Edwards, *LESBL* Logan Earth Ski Bruce Logan, *ARRA* Arbor Rail. *TNCO* Tunnel Competition, *PPMU* Powell-Peralta Mullins, *SSSC* Skull Skates Soup Can, *ALP* Alva Pig, *ALLG* Alva Logo

**Table 2** Indices for parsimony-derived trees for skateboards

| Data set | Number of trees | Consistency index | Retention index |
|---|---|---|---|
| Total | 1 | .765 | .667 |
| Slalom | 3 | .727 | .7 |
| Bowl rider | 3 | .727 | .7 |
| Freestyle | 6 | .727 | .727 |
| Street | 12 | .692 | .636 |

## 2.2 Projectile Points

Studies of projectile point evolution have sought to develop sets of characters from which to measure evolutionary process (Buchanan and Collard 2007; Darwent and O'Brien 2006; O'Brien et al. 2001; O'Brien and Lyman 2003). Projectile points are designed with haft and blade areas resulting from different design considerations and constraints (Fig. 3). Haft areas are at the proximal or base end of the artifact and are designed to facilitate connecting the stone point to a wood or bone arrow or dart shaft. There are a variety of strategies and resulting morphologies to accomplish this, and they result in forms that include lanceolate, stemmed, and notched designs (e.g., Andrefsky 2009: 179). Archaeologists generally recognize haft morphologies as carrying the most cultural information regarding style (Darwent and O'Brien 2006) even though hafting morphology may also be affected by performance factors (Hughes 1998). Blades are designed to solve functional problems, namely, piercing and cutting. However, they tend to be modified extensively through culturally and situationally variable approaches to resharpening (Flenniken and Raymond 1986). Thus, their forms can vary widely with blades short or long; margins straight, concave, or convex; and resharpening bifacial, unifacial, serrated, and beveled. Because haft and blade portions of projectile points vary due to quite different constraints, functional goals, and cultural preferences, there is a very real possibility that projectile points could evolve in a mosaic fashion. Our characters and associated character

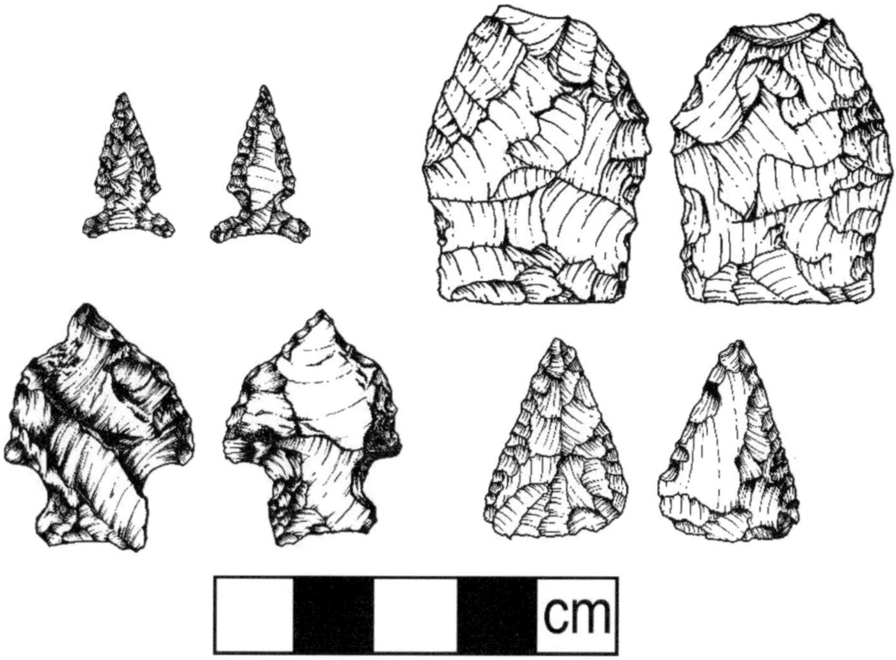


**Fig. 3** Examples of major projectile point forms used in this study (*upper left*, side-notched [II.16.2.962]; *upper right*, wide stemmed [IID.23.1.602a]; *lower left*, corner-notched/removed [IIC.23.1.238]; *lower right*, triangle [IID.1–4.1.942])

states (Table 3) partition variability in blade and haft form on projectile points using standard measures (Andrefsky 2009: 186; Darwent and O'Brien 2006).

We offer a preliminary test of mosaic evolution in projectile points using a sample of arrow points from the Bridge River housepit village in British Columbia. The Bridge River site is a large housepit village occupied most intensively by complex fisher-hunter-gatherer people during the period of ca. 1800–1100 and 500–100 years ago (Prentiss et al. 2008). Our projectile point sample derives from Housepit 20 where excavations revealed a stratified sequence of floor deposits spanning periods Bridge River (BR) 2 (1600–1300 cal. B.P.), BR 3 (1300–1100 cal. B.P.), and BR 4 (ca. 400–300 cal. B.P. in this context). Floors accumulated as occupants cyclically reroofed and re-floored their houses to eliminate periodic wood rot in roof superstructures and to eradicate insect and other pest infestations (Prentiss and Kuijt 2012). Each floor is a separate layer dominated by clay-sized particles on which household members lived, thus leaving variable quantities of artifacts, food remains, and cooking and storage features. An advantage of examining artifacts from within a single housepit occupation sequence is that despite a certain degree of similarity between artifacts in different houses, we can likely control aspects of inheritance and descent since houses were occupied by lineage groups with a largely "vertical" system of cultural inheritance (Prentiss et al. 2014). With one exception, all projectile points used here (Fig. 3) come from the older series of floors (Floors IIE and IID are from BR 2, Floors IIC–IIA are BR 3, Floor II is BR 4). The BR 4 projectile point is the only one where we have a likely break in direct cultural continuity.

**Table 3** Data used to derive projectile point study cladograms (characters and character states; data order 1–25)

| Projectile<br>Point | Data | Description |
|---|---|---|
| IIE.2.4.941 | 0101101010100100000101332 2 | Convex blade; asymmetrical corner-removed/contracting stem |
| IID.4.2.905 | 1010101010101001001110122 2 | Straight blade; split stem base, corner removed/expanding stem |
| IID.1–3.1.671 | 000010010100001000110112 21 | Short (relative to haft) straight blade; asymmetrical corner-notch/removed |
| IID.1–4.1.908 | 1001011000010011001111110 2 | Straight blade; wide stem |
| IID.1–4.1.942 | 0101011010000010001100133 2 | Straight blade, triangle shape |
| IID.23.1.602a | 0100000100000011010101323 0 | Short (relative to haft) convex blade, wide stem |
| IID.23.1.602b | 1010101010010100000110013 32 | Straight blade; side notch |
| IID.2.1.303 | 1010011110111010100011111 0 | Straight blade; side notch |
| IIC.23.1.238 | 0101100101011001010110023 1 | Short (relative to haft) concave blade; corner-notched/removed |
| IIC.1.1.62 | 0100000100000101000101112 2 | Short (relative to haft) straight/convex blade; triangle with rounded corners |
| IIB.14.3.987 | 0001010100000000010111023 0 | Concave blade, triangle shape |
| IIA.22.1.304 | 1010101010011000110100121 1 | Straight blade; side notch |
| II.16.2.962 | 1010101010101000110010121 1 | Straight blade, side notch |

Blade: blade length <20 mm (1), blade length 20–25 mm (2), blade width <15 mm (3), blade width 15–25 mm (4), point angle <50° (18), point angle 50–70° (19), blade symmetry (straight [1], concave [0], convex [3]) (23), tip convergence (24). Base: shoulder angle <90° (16), shoulder angle 90–120° (17), notch type (side or corner; 25), basal expansion (26), neck width <15 mm (5), neck width 15–20 mm (6), base length <8 mm (9), base length 8–12 mm (10), base width <12 mm (11), base width 12–16 mm (12), base shape concave (13), base shape convex (14), base shape straight (15). Blade and base (not used in independent blade and base analyses): max thickness <5 mm (7), max thickness >5 mm (8), terminations semi-abrupt (20), terminations invasive (21), retouch tep (22)

We measured variability in projectile point form and manufacture procedure with a range characters and character states (Fig. 4; Table 3). In order to measure variation in blade form, data were collected on blade length, point tip angle, blade symmetry (straight, concave, and convex), and tip convergence. Characters exclusive to base form included neck width, base length, base shoulder form (convex, concave, and straight), base shoulder angle, notch (side and corner), and basal expansion. Characters relevant to both blade and base form included maximum thickness and facial retouch pattern (invasive, semi-invasive, and step). We employed the same cladistic analysis procedures as used with skateboard decks (Table 4) and rooted all trees in the presumably oldest projectile point (deepest floor [IIE] in Housepit 20 at Bridge River).

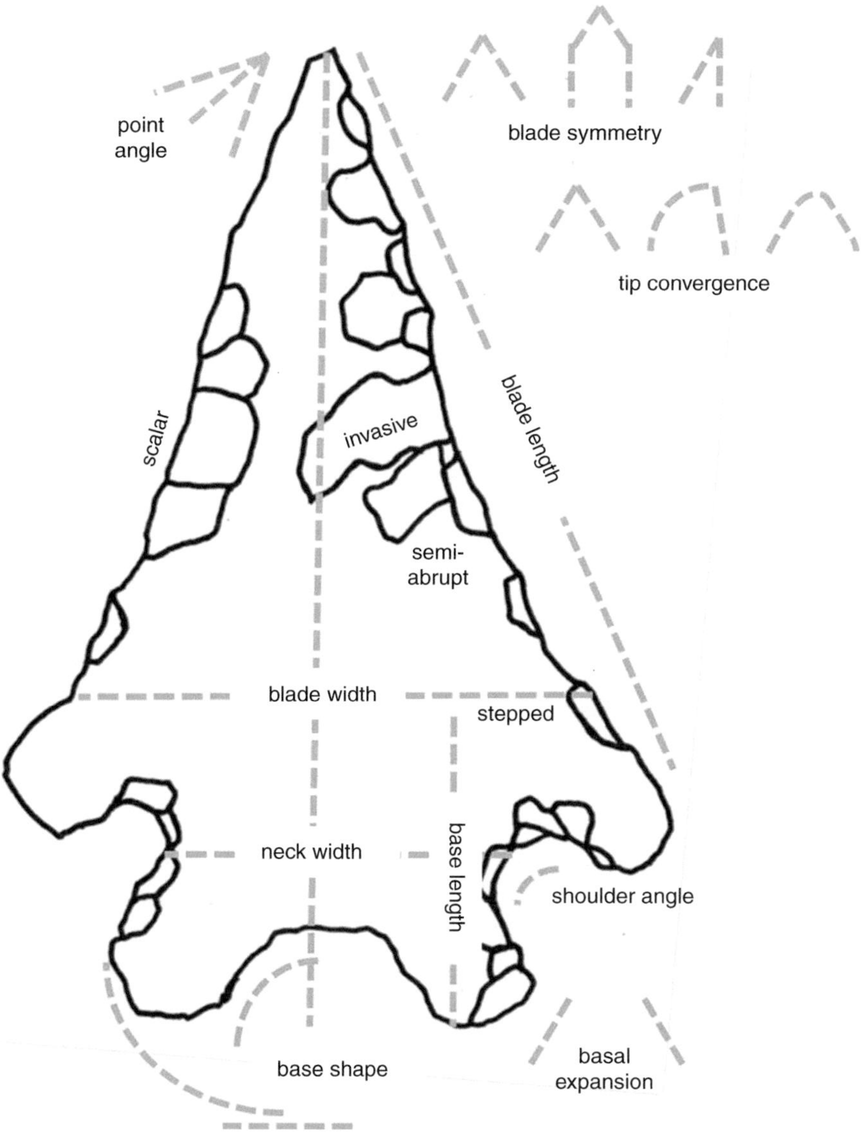


**Fig. 4** Projectile point image illustrating characters and character states used in the cladistic analysis

**Table 4** Indices for parsimony-derived trees for projectile points

| Data set | Number of trees | Consistency index | Retention index |
|---|---|---|---|
| Total | 2 | .4045 | .5319 |
| Base | 4 | .4324 | .5532 |
| Blade | 3 | .4211 | .614 |

# 3 Results

## *3.1 Skateboard Deck Results*

Analysis of the total data set resulted in a consensus cladogram with high consistency (CI) and retention (RI) indices. The consensus tree indicates three clades: slalom, bowl riders, and street/freestyle (Fig. 5). With full data the street (ARRA and SSSC) and bowl (LESBL and ALLG) boards are most highly derived (meaning most derived traits compared to root taxon – MAPE). Slalom (TNCO) is least derived. Effectively the same result was generated when variables were selectively chosen to favor slalom and bowl riding designs (Figs. 6 and 7). Results change somewhat when variables emphasize freestyle (Fig. 8) and street (Fig. 9) designs. In these cases, we produce a polytomy in each cladogram making interpretation difficult. The only obvious pattern consists of the slalom design (TNCO) as likely ancestral to all others.

The Logan Earth Ski Bruce Logan (LESBL) appears in different clades depending upon choice of variables. It appears to be a bowl rider when total, slalom, and bowl rider variables are favored. It branches closer to a freestyle board when street and freestyle variables are favored. Examining all results, street boards are always most highly derived. In contrast, slalom boards are least derived. Finally, the positions of the bowl and freestyle boards within each cladogram and phylogram change when variables shift. This helps to explain why the LESBL board actually changes clade as it embodies characteristics of both bowl rider and freestyle deck and is likely close to a common ancestor of both.

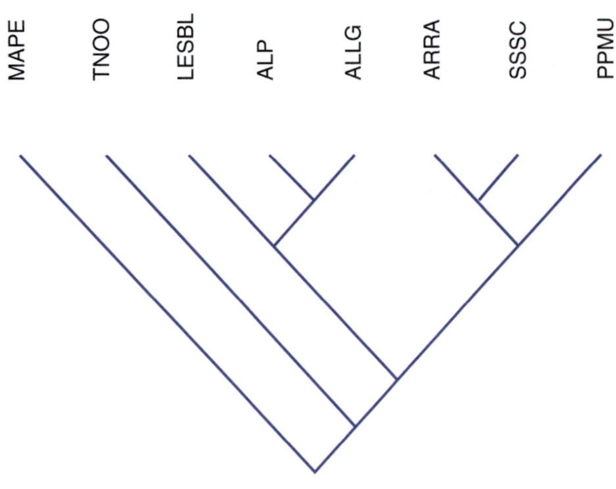


**Fig. 5** Consensus cladogram for full data set (*MAPE* Makaha Phil Edwards, *LESBL* Logan Earth Ski Bruce Logan, *ARRA* Arbor Rail. *TNCO* Tunnel Competition, *PPMU* Powell-Peralta Mullins, *SSSC* Skull Skates Soup Can, *ALP* Alva Pig, *ALLG* Alva Logo)

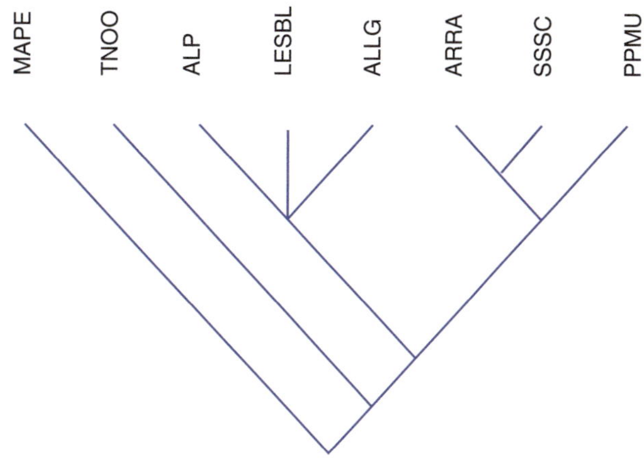


**Fig. 6** Consensus cladogram for slalom function data (*MAPE* Makaha Phil Edwards, *LESBL* Logan Earth Ski Bruce Logan, *ARRA* Arbor Rail. *TNCO* Tunnel Competition, *PPMU* Powell-Peralta Mullins, *SSSC* Skull Skates Soup Can, *ALP* Alva Pig, *ALLG* Alva Logo)

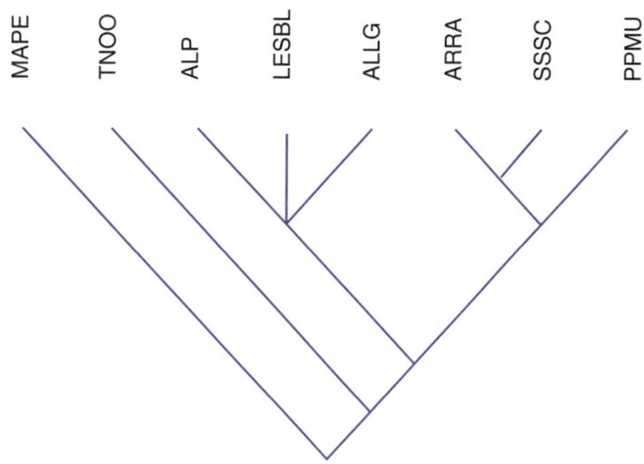


**Fig. 7** Consensus cladogram for bowl rider function data (*MAPE* Makaha Phil Edwards, *LESBL* Logan Earth Ski Bruce Logan, *ARRA* Arbor Rail. *TNCO* Tunnel Competition, *PPMU* Powell-Peralta Mullins, *SSSC* Skull Skates Soup Can, *ALP* Alva Pig, *ALLG* Alva Logo)

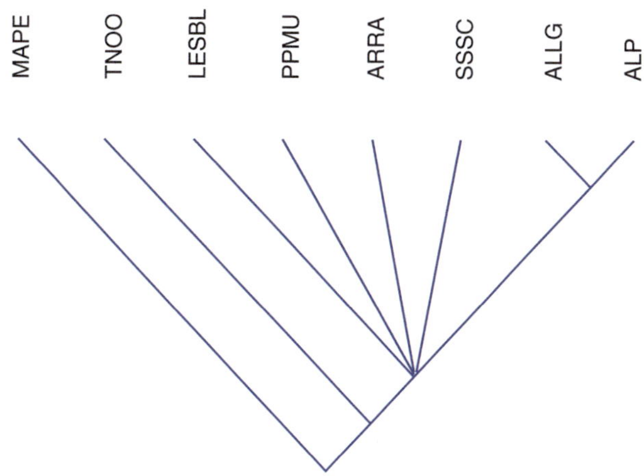


**Fig. 8** Consensus cladogram for freestyle function data (*MAPE* Makaha Phil Edwards, *LESBL* Logan Earth Ski Bruce Logan, *ARRA* Arbor Rail. *TNCO* Tunnel Competition, *PPMU* Powell-Peralta Mullins, *SSSC* Skull Skates Soup Can, *ALP* Alva Pig, *ALLG* Alva Logo)

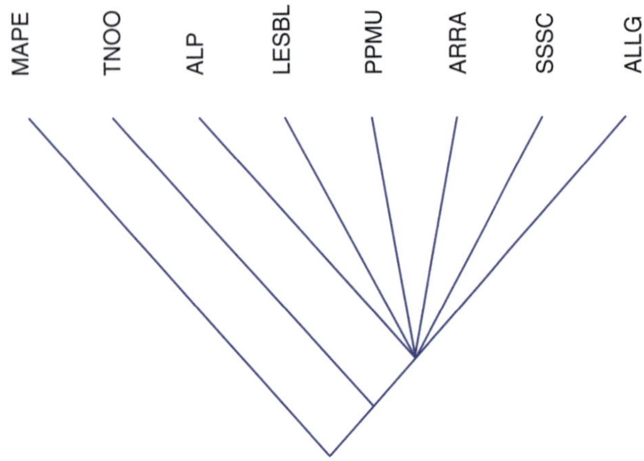


**Fig. 9** Consensus cladogram for street cruiser function data (*MAPE* Makaha Phil Edwards, *LESBL* Logan Earth Ski Bruce Logan, *ARRA* Arbor Rail. *TNCO* Tunnel Competition, *PPMU* Powell-Peralta Mullins, *SSSC* Skull Skates Soup Can, *ALP* Alva Pig, *ALLG* Alva Logo)

## 3.2 *Projectile Point Results*

The analysis of projectile point data (Table 3) also showed variation in results depending on character set. The total data set, reflecting both base and blade characters, returned two very similar maximum parsimony trees (Fig. 10) and a consensus cladogram, but comparatively somewhat lower (compared to all skateboard results) consistency (CI) and retention (RI) indices (Table 4) suggesting the possibility of a higher degree of homoplasy in the projectile point data compared to the skateboard decks. However, we recognize that low RI scores may be equally affected by other processes including variability in innovation rates (Crema et al. 2014; see also Collard et al. 2006). Multiple clade groups are recognizable and include side-notched points (e.g., IIA.22.1.304) apparently descendant from small flared base/corner-removed forms (IID.4.2.905). Corner-notched/removed points (IIC.23.1.238), triangle forms (IID.1-4.942), and wide-stem (IID.23.1.602a) variants form small clade groups and also appear descendant from the earlier corner-removed/stemmed designs. Most significantly, it is evident that points from stratum IID are foundational on every major branch. Put differently, points from strata II, IIA, IIB, and IIC generally exist as more derived twigs on the major branches. This implies that the full range of variation in projectile point style came into being during the occupation of the IID floor. Later variants represent subtle modifications of those early themes.

The projectile point base data set generated four very similar maximum parsimony trees (Fig. 11) and a consensus cladogram, along with similarly moderate CI and RI scores. The cladograms illustrate both similarities and differences to those drawn from the combination of blade and base characters. Most similar is retention of the side-notch point clade (e.g., IIA.22.1.304) with its likely ancestor, the small flared base/corner-removed form (e.g., IID.4.2.905). Also similar is the distinct corner-notched/removed point clade (e.g., IIC.23.1.238). Different from the total data results is the lack of coherent clade formation between wide-stemmed and triangular points. This is likely because the base form is a little different between the two. Distinctions only emerge at the midline and toward the distal tip. The base

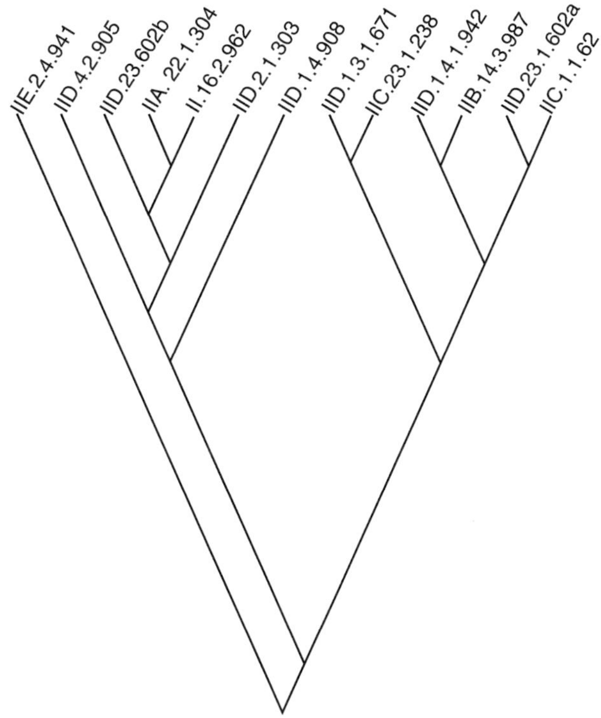


**Fig. 10** Maximum parsimony tree (first of two) for projectile point blade and base data

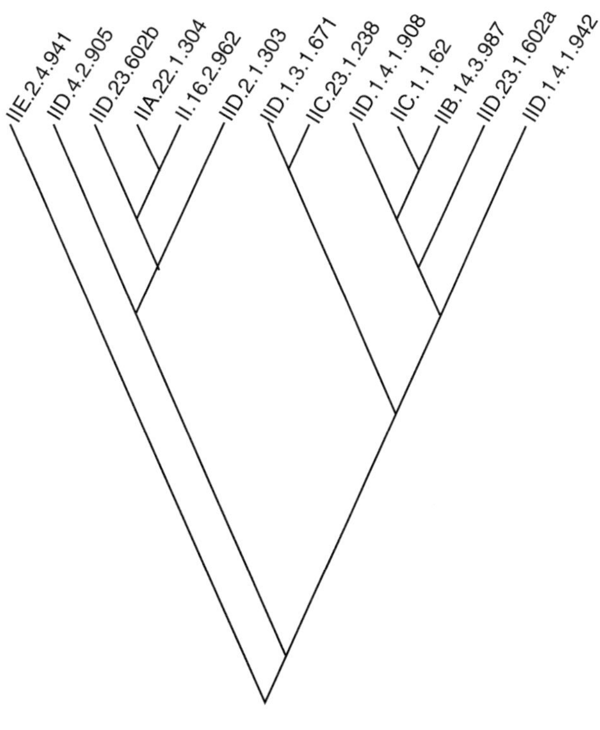


**Fig. 11** Maximum parsimony tree (second of four) for projectile point base data

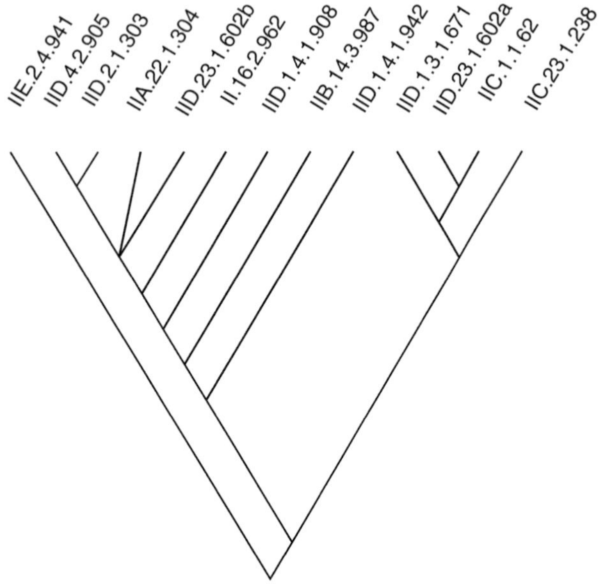


**Fig. 12** Consensus cladogram for projectile point blade data

results also continue to support the conclusion that maximum diversity of styles developed on the IID occupation floor.

The blade-oriented data set produced three similar maximum parsimony trees, a consensus tree (Fig. 12), and, again, moderately low CI and RI scores. Two clade groups are evident. The larger (left side of Fig. 12) is dominated by side-notch and triangle forms, along with one wide stem somewhat resembling a triangle (IID.1-4.1.908). The other clade consists of corner-notched/removed, wide-stem, and one triangular form with ovoid corners (IIC.1.1.62). This result is somewhat different from trees derived from total data and bases forms. The primary criterion distinguishing members of each clade appears to be blade symmetry, point angle, and tip form, likely responding most significantly to variation in episodes of resharpening, thus not inherited stylistic conventions. This suggests that the logic behind manufacture activities associated with blades was not the same as that of hafts. Interestingly, however, as noted with regard to total and haft-oriented data sets, it is still characterized by the consistent presence of older points (floor IID) in less derived contexts than those of later floors (II, IIA, IIB, and IIC).

## 4 Discussion

By about the mid-1970s (the time of the LESBL skateboard deck), divergent skater needs began to put pressure on makers to take deck designs in two new directions. One was toward longer and wider boards with kicktails and short noses (the bowl riders) and the other was toward short narrow bodies (much like earlier boards) but with long noses and kicktails (the freestyle boards). This pressure was not on overall morphology but select attributes. Some basic design elements did not change – general shape – wider on tail than nose, max width at front-center, and kicktail. This implies that mosaic evolution did play a major role in skateboard evolution. A major implication is that analyses of prehistoric artifact histories could be heavily affected by our choice of variables.

Evolutionary archaeologists have focused on projectile points to illustrate uses of cladistic methodology (O'Brien et al. 2001; O'Brien and Lyman 2003) to address patterns of land use and colonization (Buchanan and Collard 2007) and to develop models of ancient technological evolution (Darwent and O'Brien 2006). Often large data bases are used to derive complex cladograms based upon variables associated with two areas of divergent morphology: blade and haft area. While it is well known that differential requirements and constraints impact manufacturing decisions associated with the different portions of the artifact, analysts have consistently combined characters from both areas within single studies (e.g., Buchanan and Collard 2007; O'Brien and Lyman 2003). With many variables impacting projectile point form, it is clear that evolutionary process has significant potential to occur in a mosaic fashion and, thus, confound our attempts at creating accurate trees without careful consideration of variables.

Our analysis of projectile points from Housepit 20 at the Bridge River site offers a number of implications. First, while the results of our study are clearly impacted by homoplasy or other confounding processes, we do recognize a consistent and coherent branching pattern. Projectile points from the second oldest floor, Stratum IID, appear on every branch, implying that the nearly full range of variation appeared early in the history of this house. Second, details of clade membership are affected by choice of characters whether base/haft or blade oriented. It should not be surprising that the major distinctions between groups of points, when measured using blade characteristics, are likely the result of differential resharpening activities. Given the fact that all data sets indicate an early pattern of branching, we can tentatively conclude that some degree of descent with modification did occur and that limited mosaic evolution occurred in haft forms. Resharpening of blades in this case was likely situationally variant and thus probably not associated with cultural inheritance. Even if the Bridge River projectile point blades were modified to solve situational contingencies, we could still expect examples elsewhere whereby blade form did evolve in a modular fashion. There is significant blade variation, for example, between projectile points with similar hafting morphologies associated with the Cody Complex and Western Stemmed traditions of Northwestern North America (Chatters et al. 2012; Larson 2012).

A final implication of this study concerns evolutionary entities. Decades ago, Dunnell (1989) cautioned evolutionary archaeologists to be clear in defining the basic unit of evolution. The lesson of mosaic evolution is that the definition may not be clear without careful consideration of attributes. Research in modular evolution (e.g., Schlosser 2002) suggests that not only do we need to be careful in our consideration of evolutionary targets but that we will not fully understand the nature and process of mosaic evolution without understanding the linkages between modules and larger wholes and their potential for recombination over time into novel configurations. For skateboards, it meant a variety of modifications to the basic design during a period of significant branching that was followed by a later period marked by blending of those disparate design elements to create even more unique designs. In effect, the core design changed little while specific modules changed substantially.

Our study of Bridge River projectile points offers an example of how projectile points could potentially evolve in a modular fashion. Our trees derived from the combined base and blade form data set actually differed relatively little from the base-only trees. An implication is that data derived from blade form will not necessarily confound an analysis that primarily emphasizes haft area form. However, this does not mean that analysts should not be aware that variation in blade form can be substantially modified for reasons unrelated to the logic behind overall point form, especially haft area form. Flenniken and Raymond's (1986) cautions regarding the effects of variability in projectile point manufacture and resharpening decisions are mirrored in our data. Even if our results are affected by decisions that are unrelated to inherited logic, they open the possibility that such phenomena could occur (e.g., in certain Paleo-Indian point traditions). Shott (2011) argued that lithic technologists are still far from developing a comprehensive theory of projectile point evolution suggesting that analysts need to be concerned with (among other things) a range of factors such as raw material, production process, and use-life. We suggest that an understanding of modularity of projectile point evolution could be a useful concept in providing us with a means to better understand divergent effects of evolutionary forces on these tool forms.

**Acknowledgements** We thank Nathalie Gontier for inviting us to participate in the AAA session in San Francisco and to contribute this paper. Bridge River projectile point data derive from excavations funded by the National Science Foundation (BCS-0713013) and conducted via a collaborative partnership between the University of Montana and the Bridge River Band (Xwisten) and St'át'imc Nation. Skateboards in Fig. 1 were drawn by Eric Carlson. We thank Larissa Mendoza Straffon and two peer reviewers for their useful comments.

## References

Andrefsky, W., Jr. (2009). *Lithics: Macroscopic approaches to analysis*. Cambridge: Cambridge University Press.

Boyd, R., Mulder, M. B., Durham, W. H., & Richerson, P. J. (1997). Are cultural phylogenies possible? In P. Weingart, S. D. Mitchell, P. J. Richerson, S. Maasen (Eds.), *Human by nature: Between biology and the social sciences* (pp. 355–384). Mahwah: Lawrence Erlbaum Associates, Publishers.

Brace, C. L. (1995). Biocultural interaction and the mechanism of mosaic evolution in the emergence of "Modern" morphology. *American Anthropologist*, *97*, 711–721.

Brandon, R. N. (1999). The units of selection revisited: The modules of selection. *Biology and Philosophy*, *14*, 167–180.

Buchanan, B., & Collard, M. (2007). Investigating the peopling of North America through cladistic analysis of early paleoindian projectile points. *Journal of Archaeological Science*, *26*, 366–393.

Chatters, J. C. (2009). A macroevolutionary perspective on the archaeological record of North America. In A. M. Prentiss, I. Kuijt, & J. C. Chatters (Eds.), *Macroevolution in human prehistory: Evolutionary theory and processual archaeology* (pp. 213–234). New York: Springer.

Chatters, J. C., Hackenberger, S., Lenz, B., Prentiss, A. M., Thomas, J-L. (2012). The paleoindian to archaic transition in the pacific northwest: In situ development or ethnic replacement. In C. B. Bousman, & B. J. Vierra (Eds.), *On the brink: Transformations in human organization and adaptation at the pleistocene-holocene boundary in North America* (pp. 37–66). Texas: Texas A & M Press, College Station.

Clarke, J. A., & Middleton, K. M. (2008). Mosaicism, modules, and the evolution of birds: Results from a Bayesian approach to the study of morphological evolution using discrete character data. *Systematic Biology*, *57*, 185–201.

Collard, M., Shennan, S. J., & Tehrani, J. J. (2006). Branching, blending, and the evolution of cultural similarities and differences among human populations. *Evolution and Human Behavior*, *27*, 169–184.

Crema, E. R., Kerig, T., & Shennan, S. (2014). Culture, space, and metapopulation: A simulation-based study for evaluating signals for blending and branching. *Journal of Archaeological Science*, *43*, 289–298.

Dagg, J. L. (2011). Exploring mouse trap history. *Evolution: Education and Outreach*, *4*, 397–414.

Darwent, J., & O'Brien, M. J. (2006). Using cladistics to construct lineages of projectile points from Northeast Missouri. In C. P. Lipo, M. J. O'Brien, M. Collard, & S. J. Shennan (Eds.), *Mapping our ancestors: Phylogenetic approaches in anthropology and prehistory* (pp. 185–208). New York: Aldine Transaction.

Dunnell, R. C. (1989). Aspects of the application of evolutionary theory in archaeology. In C. C. Lamberg-Karlovsky (Ed.), *Archaeological thought in America* (pp. 35–49). Cambridge: Cambridge University Press.

Eldredge, N. (2000). Biological and material cultural evolution: Are there any true parallels? *Perspectives in Ethology*, *13*, 113–153.

Eldredge, N. (2006). Forward. In C. Lipo, M. J. O'Brien, M. Collard, & S. J. Shennan (Eds.), *Mapping our ancestors: Phylogenetic approaches in anthropology and prehistory* (pp. xiii–xvi). New Brunswick: Aldine Transaction.

Eldredge, N. (2009). Material cultural macroevolution. In A. M. Prentiss, I. Kuijt, & J. C. Chatters (Eds.), *Macroevolution in human prehistory: Evolutionary theory and processual archaeology* (pp. 297–316). New York: Springer.

Eldredge, Niles (2011) Paleontology and Cornets: Thoughts on Material Cultural Evolution. Evolution: Education and Outreach 4:364-373.

Flenniken, J. J., & Raymond, A. W. (1986). Morphological projectile point typology: Replication experimentation and technological analysis. *American Antiquity*, *51*, 603–614.

Goodale, N., Jones, G. T., & Beck, C. (2011). Natural selection and material culture. *Evolution: Education and Outreach*, *4*, 427–434.

Goodrich, J. (2010). Skateboard history time line – 1920s to 2010. http://jimgoodrich.net/skateboardtimeline.htm.

Gould, S. J. (1977). *Ontogeny and Phylogeny*. Cambridge: Belknap Press.

Hammer, Ø., Harper, D. A. T., Ryan, P. D. (2001). PAST: Paleontological statistics software package for education and data analysis. *Palaeontologia Electronica*, 4(1), 9. http://palaeo-electronica.org/2001_1/past/issue1_01.htm.

Hughes, S. S. (1998). Getting to the point: Evolutionary change in prehistory weaponry. *Journal of Archaeological Method and Theory*, *5*, 345–408.

Jordan, P., & Shennan, S. (2009). Diversity in hunter-gatherer technological traditions: Mapping trajectories of cultural 'descent with modification' in Northeast California. *Journal of Anthropological Archaeology*, *28*, 342–365.

Kemp, T. S. (2007a). The concept of correlated progression as the basis of a model for the evolutionary origin of major new taxa. *Proceedings of the Royal Society B*, *274*, 1667–1673.

Kemp, T. S. (2007b). The origin of higher taxa: Macroevolutionary processes, and the case of the mammals. *Acta Zoologica*, *88*, 3–22.

Larson, Mary Lou (2012) The Paleoindian to Archaic Transition: The Northwestern Plains and Rocky Mountains. In C. Britt Bousman and Bradley J. Vierra (eds.), From the Pleistocene to the Holocene: Human Organization and Cultural Transformations in Prehistoric North America, pp. 149-170. Texas A&M University Press, College Station.

Lü, J., Unwin, D. M., Jin, X., Liu, Y., & Ji, Q. (2010). Evidence for modular evolution in a long-tailed pterosaur with a pterodactyloid skull. *Proceedings of the Royal Society B*, *277*, 383–389.

Maglio, V. J. (1973). Origin and evolution of the elephantidae. *Transactions of the American Philosophical Society*, *63*, 1–149.

McHenry, H. H. (1994). Tempo and mode in human evolution. *Proceedings of the National Academy of Science*, *91*, 6780–6786.

O'Brien, M. J., & Lyman, R. L. (2003). *Cladistics and archaeology*. Salt Lake City: University of Utah Press.

O'Brien, M. J., Darwent, J., & Lyman, R. L. (2001). Cladistics is useful for reconstructing archaeological phylogenies: Paleoindian points from the Southeastern United States. *Journal of Archaeological Science*, *28*, 1115–1136.

Pääbo, S. (2003). The mosaic that is our genome. *Nature*, *421*, 409–412.

Prentiss, A. M., & Kuijt, I. (2012). *People of the middle fraser canyon: An archaeological history*. Vancouver: University of British Columbia Press.

Prentiss, A. M. Cross, G., Foor, T. A., Markle, D., Hogan, M., & Clarke, D. S. (2008). Evolution of a late prehistoric winter village on the interior plateau of British Columbia: Geophysical investigations, radiocarbon dating, and spatial analysis of the bridge river site. *American Antiquity*, *73*, 59–82.

Prentiss, A. M., Goodale, N. B., Harris, L. E., & Crossland, N. (2014). The evolution of the ground slate tool industry at the bridge river site, British Columbia. In N. B. Goodale, & W. Andrefsky Jr. (Eds.), *Lithic technological systems: Stone, evolution, and behavior*. Cambridge: Cambridge University Press, (in press).

Prentiss, A. M., Skelton, R. R., Eldredge, N., & Quinn, C. P. (2011). Get rad! The evolution of the skateboard deck. *Evolution: Education and Outreach*, *4*, 379–389.

Richerson, P. J., & Boyd, R. (2005). *Not by genes alone: How culture transformed human evolution*. Chicago: University of Chicago Press.

Schlosser, G. (2002). Modularity and the units of evolution. *Theory in Bioscience*, *121*, 1–80.

Shott, M. J. (2011). History written in stone: Evolutionary analysis of stone tools in archaeology. *Evolution: Education and Outreach*, *4*, 435–445.

Skelton, R. R., & McHenry, H. M. (1998). Trait list bias and a reappraisal of early hominid phylogeny. *Journal of Human Evolution*, *34*, 109–113.

Stanley, S. M. (1998). *Macroevolution: pattern and process*. Baltimore: Johns Hopkins University Press.

Tehrani, J. (2011). Patterns of evolution in Iranian tribal textiles. *Evolution: Education and Outreach*, *4*, 39–396.

Tehrani, J., & Collard, M. (2002). Investigating cultural evolution through biological phylogenetic analyses of Turkmen textiles. *Journal of Anthropological Archaeology*, *21*, 443–463.

Weyland, J. (2002). *The answer is never: A Skateboarder's History of the World*. New York: Grove Press.

# Mind the Network: Rock Art, Cultural Transmission, and Mutual Information

**Inés Caridi and Vivian Scheinsohn**

**Abstract** Decorative patterns have long been considered suitable for determining descent, since they are categorized as homologous and adaptively neutral. Rock art, for its part, has often been left aside due to a lack of chronological control. In this paper, we propose a way to treat rock art in order to track Cultural Transmission Paths by means of motif distribution using Northwestern Patagonia as a case study. We present a theoretical and methodological framework for modeling Cultural Transmission Archaeological Paths by constructing a Mutual Information Network between motifs, identifying clusters and defining their associated Site Networks. The results allow us to suggest a hypothetical nuclear region, well known and transited by hunter-gatherers, with few connections to the more distant parts of the study area. This pattern may be related to Patagonia's population models and fit the suggestion from other fields of inquiry that a sparsely connected and not unnecessarily complex network will be robust enough to sustain information flux.

**Keywords** Rock Art • Patagonia • Hunter gatherers • Mutual Information

## 1 Introduction

Since Dunnell's seminal work (Dunnell 1978), several researchers have considered stylistic characters as adaptively neutral. Decorative patterns, in particular, have been considered as nonfunctional or selectively neutral, since they are not

I. Caridi (✉)
Institute of Calculus, School of Exact and Natural Sciences,
University of Buenos Aires, National Scientific and Technical Research Council,
Buenos Aires, Argentina
e-mail: ines@df.uba.ar

V. Scheinsohn
Department of Anthropology, National Institute of Anthropology and
Latin American Thought, National Scientific and Technical Research Council,
University of Buenos Aires, Buenos Aires, Argentina
e-mail: scheinso@retina.ar



L. Mendoza Straffon (ed.), *Cultural Phylogenetics: Concepts and Applications in Archaeology*, Interdisciplinary Evolution Research,
DOI 10.1007/978-3-319-25928-4_7

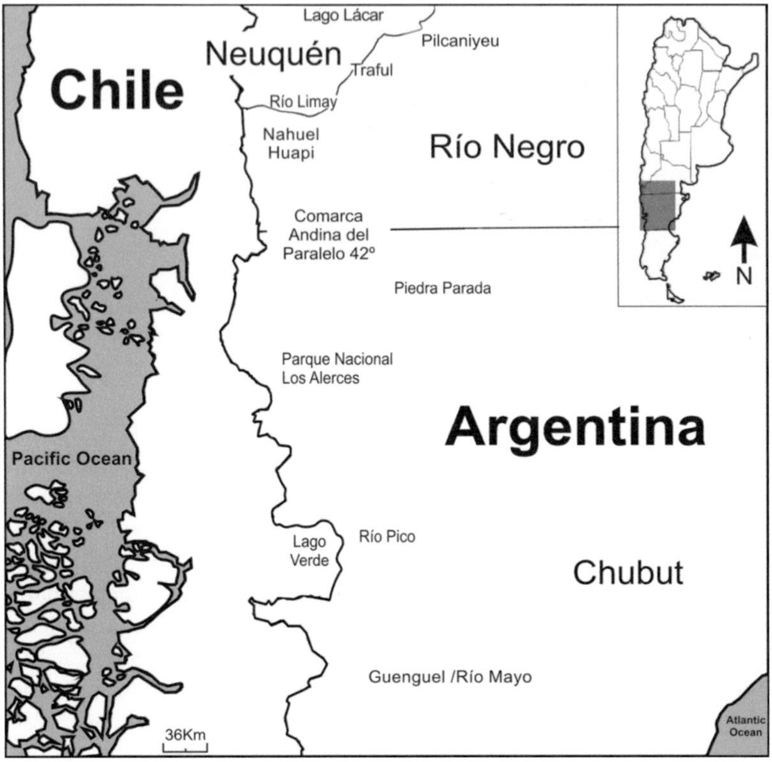


**Fig. 1** Map of the study area of Patagonia

tied to functional constraints. For instance, in pottery, decorative patterns are considered so complex that the probability of duplication by chance would be small. So, if two vessels with the same decorative pattern are found "(…), the more parsimonious explanation of such phenomenon is that the vessels share a common developmental history and are from the same tradition" (O'Brien and Lyman 2003: 19). This property of decorative patterns has allowed the establishment of cultural lineages derived from them. There are, for example, several works that have employed ceramic decorative patterns to establish cultural lineages (i.e., Neiman 1995; Shennan and Wilkinson 2001; Cochrane and Lipo 2010). The same has been done with textiles (Tehrani and Collard 2002) and basketry (Jordan and Shennan 2003). However, rock art, despite its potential to track Cultural Transmission processes and their patterns, has been left aside. In this work we present a theoretical and methodological framework for modeling Cultural Transmission Archaeological Paths by constructing a Mutual Information Network between motifs, identifying clusters, and defining their associated Site Networks. We will develop this proposal applying it to our study area, NW Patagonia, which includes nine regions (see Fig. 1). Table 1 shows a detailed list of sites by region.

**Table 1** Archaeological sites in our study area, organized in 9 regions (first column)

| Region | Site | Simplified site name | Source | Amount of unrepeated motifs |
|---|---|---|---|---|
| Traful | Alero Las Mellizas | ALM | Silveira and Fernández (1991) | 17 |
| | Alero Los Cipreses | ALC | Silveira and Fernández (1991), Silveira (1996) | 5 |
| | Alero Lariviere | AL | Silveira (1988–1989, 1999), Silveira and Fernández (1991) | 10 |
| Piedra Parada | Campo Moncada 1 | CM1 | Aschero et al. (1983), Onetto (1986–1987), Pérez de Micou (1979–1882) | 7 |
| | Campo Nassif 1 | CN1 | Aschero et al. (1983), Onetto (1986–1987), Pérez de Micou (1979–1882) | 14 |
| | Piedra Parada 1 | PP 1 | Aschero et al. (1983), Onetto (1986–1987), Pérez de Micou (1979–1982) | 12 |
| | Piedra Parada 4 | PP 4 | Aschero et al. (1983), Onetto (1986–1987), Pérez de Micou (1979–1982) | 12 |
| | Campo Cretón 1 | CCR1 | Aschero et al. (1983), Onetto (1986–1987), Pérez de Micou (1979–1982) | 11 |
| Comarca Andina del Paralelo 42° | Peñasco | PE | Firsthand data set | 16 |
| | Cerro Pintado | CP | Firsthand data set | 29 |
| | Risco de Azócar 1 | RA1 | Firsthand data set | 12 |
| | Risco de Azócar 2 | RA2 | Firsthand data set | 13 |
| | El Radal | ER | Firsthand data set | 15 |
| | Paredón Lanfré | PL | Firsthand data set | 12 |
| | Peumayén 2 | PEU2 | Podestá et al. (2009) | 17 |
| Pilcaniyeu | Cueva Sarita 3 | CS3 | Boschín (2000, 2009) | 13 |
| | Cueva Sarita 4 | CS4 | Boschín (2000, 2009) | 1 |
| | Cueva Comallo 1 | CCO1 | Boschín (2000, 2009) | 16 |
| | La Figura 1 | LF1 | Boschín (2000, 2009) | 2 |
| | Cueva Pulpulcurá 2 | PUL2 | Boschín (2000, 2009) | 5 |
| | Abrigo de Pilcaniyeu | PIL | Boschín (2000, 2009) | 8 |
| | Cueva Cuadro Leleque 1 | CCLE1 | Boschín (2000, 2009) | 15 |
| | Cueva 1 del río Pichileufu | PICH | Boschín (2000, 2009) | 10 |
| | IV 2a Puerto Tranquilo Sección 17 | PT | Pedersen (1978) | 16 |

(continued)

**Table 1** (continued)

| Region | Site | Simplified site name | Source | Amount of unrepeated motifs |
|---|---|---|---|---|
| Nahuel Huapi | Cerro Leones | CLEO | Vignati (1944) | 4 |
| | LNH1 Puerto Tigre | Pti | Pedersen (1978) | 20 |
| | LNH2 Nariz del Diablo | ND 1 | Pedersen (1978) | 13 |
| | IV 4 Puerto Vargas | PV | Pedersen (1978) | 10 |
| | Estancia Huemul | EHUE | Vignati (1944) | 5 |
| | IV 3 al Norte de Puerto Vargas | NPV | Pedersen (1978) | 7 |
| | Chavol I | CHAI | Albornoz (1996) | 1 |
| | Chavol II | CHAII | Albornoz (1996) | 7 |
| | Bahia Lopez | BL | Albornoz (1996) | 6 |
| | Cerro Campanario I | CCAM | Albornoz (1996) | 6 |
| | El Trébol | ET | Albornoz (1996), Hajduk et al. (2004, 2009) | 11 |
| Parque Nacional Los Alerces | Alero del Sendero de Interpretación | ASI | Arrigoni (1997), Arrigoni and Fernández (2004) | 12 |
| | Alero del Shamán | ASH | Arrigoni (1997) Arrigoni and Fernández (2004) | 19 |
| Lago Lácar | Catritre 1 | CA1 | Albornoz and Cúneo (2000) | 13 |
| | Quila Quina 1 | QQ1 | Albornoz and Cúneo (2000) | 13 |
| | Curruhuinca 1 | CUR1 | Albornoz and Cúneo (2000) | 10 |
| Guenguel/Río Mayo | Guenguel | GUE | Pérez de Micou et al. (2009) | 5 |
| | Manantial 1 | MA1 | Pérez de Micou et al. (2009) | 5 |
| | Manantial 2 | MA2 | Pérez de Micou et al. (2009) | 7 |
| | Viejo Corral | VCO | Pérez de Micou et al. (2009) | 7 |
| | Bardas Blancas | BB | Pérez de Micou et al. (2009) | 10 |
| Lago Verde/Río Pico | Acevedo 1 | A1 | Firsthand data set | 9 |
| | Solís 1 | S1 | Firsthand data set | 5 |
| | Lago Verde 1 | LV1 | Firsthand data set | 3 |
| | Lago Verde 2 | LV2 | Firsthand data set | 2 |

We present the name of the site (second column); their simplified names (third column) source of information for each site (fourth column) and amount of motifs for each site (fifth column)

## 2 Rock Art and Information

Rock art is one of humankind's most ancient channels of visual communication. In fact, among prehistoric hunter-gatherers, few others existed beside face-to-face interaction (e.g., smoke signals, stylistic messages conveyed in artifacts, etc.; see Wobst 1977). Archaeologists have always been aware of the communicative role of rock art and its information storing capacity. This idea has been formalized in the "information storage model" (Barton et al. 1994; Conkey 1978; Gamble 1991; Mithen 1988; among others), which argues that rock art can store different kinds of information: from social interaction (e.g., Conkey 1978; Gamble 1991) to potential resources (e.g., Mithen 1988) or altered states of consciousness ("shamanic approach"; see Dowson 1998; Lewis-Williams and Dowson 1988). Whallon has argued that such an information storage system only functions as long as the knowledge of how to retrieve that information is present in a social group. When that knowledge is lost, it becomes impossible to access it (Whallon 2011). Nevertheless, setting meaning aside (which we will not discuss in this work, as explained below), there is information contained in the spatial distribution of rock art motifs, which could be retrieved. We will specifically focus on this point; appealing to Information Theory (Whallon 2011 is also anticipated this possibility).

## 3 Information Theory and Mutual Information

As Mackay (2003) has pointed out, communication does not have to involve information transfer from one point to another, as in a phone call. When we write a file on a computer and save it, for example, we can retrieve it from the same location, but at a later time. This was recognized by Wobst who, in his 1977 pioneering paper, affirmed that emission and reception may be separated from each other spatially and temporally. That is what happens when we visit a rock art site: the message is transmitted from the past (recent or remote) to the present. In fact, any temporal process could be thought of as a communication channel that links the past to the future (Crutchfield et al. 2009). But, in the case of rock art, we also deal with a spatial dimension set across a geographical area. Therefore, the spatial distribution of rock art motifs, paraphrasing White's affirmation about artifacts, can be viewed as the material residues of "spatially-situated, network-mediated systems of social learning and information exchange" (White 2013:1).

As mentioned above, Information Theory allows us to treat information content without any concern for meaning. When we observe two sources of messages, we are interested in detecting whether there is any type of correlation between the messages sent from both of the sources. Mutual Information is a measure of the amount of information that one message contains about the other. It measures the average reduction in uncertainty about a message from the first source that results from knowing the message from the second source, or vice versa (see details below in Appendix 1). What we want to stress here is that the Mutual Information concept (Cover and Thomas 1991), as applied in communications, could play an important role when analyzing rock art. Specifically, we propose applying Mutual Information to formalize correlations between presence and absence of rock art motifs in archaeological sites. Mutual Information (see Appendix 1 for details) quantifies a correlation, describing a pattern of motif presence and absence across different sites. But, what process produces that pattern? Given a motif is complex enough, and assuming a correlation threshold that allows us to discard chance as a factor, if a correlation exists, we may assume that the only process that could explain this pattern is Cultural Transmission, no matter the social (individual or group), temporal (motifs inscribed at the same time or in different moments), or spatial scales (in a close or wide range) at which occurs. Rock art, as a communicational channel, is intended to transmit information to others, not necessarily present at the moment when a motif is made. In this sense, it could function as either an inter- or intragenerational channel.

As an example of how to apply Mutual Information to rock art, let us introduce variable $X$, representing a particular motif, which can take two possible values: 0 means the absence of the motif in a particular site and 1 its presence. The same occurs with a second motif (variable $Y$). Mutual Information between these two motifs, $I(X,Y)$, quantifies their correlation: greater values of $I(X,Y)$ means that the presence (or absence) of motif $X$ occurs simultaneously with the presence (or absence) of motif $Y$.

To visualize the patterns of Mutual Information, we will construct a network defined by two sets: a set of nodes (rock art motifs) and a set of links (Mutual Information between them). In a social network, nodes can represent individuals, and links can represent some type of relationship or cooperation among individuals (see Watts and Strogatz 1998). In biology, networks are used to represent patterns of interaction between biological elements; sometimes, links in a network represent explicit relations between nodes, because there is a physical element such as a route or because there is a known interaction. But sometimes connections are not explicit. In such cases, connections may be inferred from the attributes of the nodes, by formalizing a correlation network. In genome research, gene coexpression networks were formalized by detecting strong correlations that may exist between gene expression patterns (Torkamaniet et al. 2010). In this area, Mutual Information is used to detect correlations in order to formalize the Mutual Information Network of coexpression of proteins (Simonetti et al. 2013). In our case, links will represent certain values of Mutual Information between pairs of motifs. If this Mutual Information Network (MIN) can help us to formalize motif correlations, we also need to evaluate how this MIN was established in a certain space, i.e., our study area. In so doing we will determine the Cultural Transmission Paths that constituted the MIN.

## 4 Cultural Transmission, Rock Art, and Archaeological Cultural Transmission Paths

### *4.1 Cultural Transmission Theory in Systemic and Archaeological Contexts*

Cultural Transmission could be defined as the "information capable of affecting individuals' behavior that they acquire from other members of their species through teaching, imitation, and other forms of social transmission" (Richerson and Boyd 2005: 5). The debate around Cultural Transmission Theory in anthropology (Cavalli-Sforza and Feldman 1981; Boyd and Richerson 1985) begun with the inception of what has been termed Dual-inheritance or Gene-culture coevolutionary models (GCC models; see Ross and Richerson 2014) in human evolution. These models posit that although genetic evolution is still present in human evolution, there are several other forces driving cultural evolution that are distinctive and derived from the fact that culture can be transmitted through social networks in ways that are much more complex than gene transmission. Cultural agents can often choose from a wide array of cultural variants, and the choice could be random or nonrandom. In the last case, biasing forces could be involved, such as direct, frequency, prestige, or other biases (Boyd and Richerson 1985). These biases and nonrandom innovation can create strong directional forces in cultural evolution (Ross and Richerson 2014: 103–104). Non-genetic transmission of behavior is much more varied and flexible than genetic transmission (Barton and Clark 1997): Cultural Transmission can go from parents to offspring (vertical transmission) as in genetic transmission but also can be dispersed among non-related individuals (horizontal transmission; see Borgerhoff Mulder et al. 2006). Current approaches to studying Cultural Transmission are set either at a micro- or macro temporal and spatial scales of analysis (Stark et al. 2008). While Cultural Transmission is studied among living populations, archaeological studies "look at actual, unsimulated, large-scale transmission through entire populations involving countless unidentified individuals from successive biological generations" (Mesoudi 2008: 97). As Mesoudi has stated, "the macroevolutionary patterns studied by archaeologists are the product of the microevolutionary transmission mechanisms studied by psychologists but the gap between these two scales of analysis still remains unexplored" (Mesoudi 2008: 99). Although the work developed around Cultural Transmission is quite abundant, few papers have focused on the networks that allow that transmission. Most works are based on the assumption that information transmission is unstructured although "the structure of interaction makes a difference to the outcomes of Cultural Transmission processes and should be considered a potentially important causal factor" (White 2013: 22).

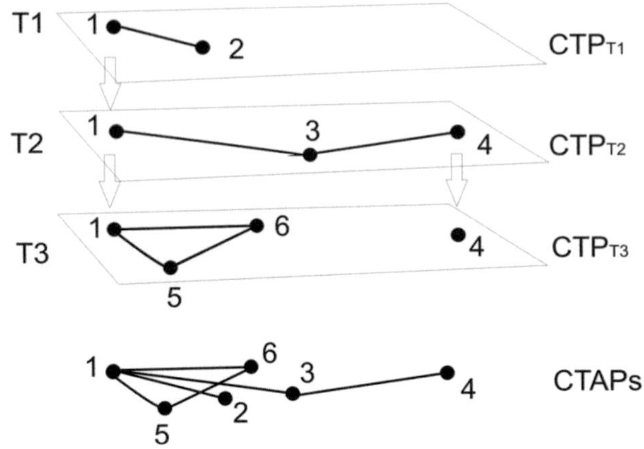


**Fig. 2** Cultural Transmission Paths and the construction of a Cultural Transmission Archaeological Path

Rock art is a product of Cultural Transmission. In a systemic context (*sensu* Schiffer 1972), we assume that the process leading to the distribution of motifs in the landscape begins when someone paints a motif on one site. That very same person, or another, could store it in his/her memory and reproduce it at the same site or elsewhere (immediately or at a later moment). There, the motif is seen by others who repeat it and can add other motifs. This process iterates for days, years, and/or centuries. As a result, some motifs will not be reproduced, while others will be distributed in a wide area. In this manner, we obtain a process of Cultural Transmission in which a social network replicates a set of motifs. The product of that process is a differential pattern of motif distribution in the landscape. In archaeology, we cannot track that social network because it is gone, but we have a "fossil" pattern, a relic of that process. Since the pattern does not mirror the network that produced it (reasons below), we need to introduce a new concept in order to separate the process (related to social network activities) and its (patterned) material evidence. On this basis, we can model a Cultural Transmission Path (CTP), a pattern left as a rock art motif distribution in the landscape (space) corresponding to a social network (social relationships) and a certain moment (time).

Figure 2 further explains this concept. At time 1 (T1), there were two nodes (sites), 1 and 2, in a given region (represented by a rectangle). As a result of a social network, there was information flow between those sites (represented by rock art motif sharing, similar vessel decoration, or else). In order to model it, we establish a link between those sites. This model is a Cultural Transmission Path for T1 ($CTP_{T1}$). At time 2 (T2), node 1 is still active but node 2 is inactive. Instead, two other nodes arise, 3 and 4. For the same reasons as before, we establish a link between 1 and 3 and between 3 and 4. So we now have the Cultural Transmission Path for T2 ($CTP_{T2}$) which is different from $CTP_{T1}$, although they share a node (1). At time 3 (T3) we can see that node 1 is still active, but its Cultural Transmission Path ($CTP_{T3}$) relates it to two new nodes, 5 and 6. Node 3 is inactive, while node 4 is still active, but it does not share a CTP with node 1 anymore.

From these three situations, archaeologically, we only can establish the Cultural Transmission Archaeological Paths (CTAPs) that are represented in Fig. 2 below. CTAPs are not related to a specific time, but they are the sum of many CTP. We prefer the term CTAPs for the latter since what we have is an archaeologically determined channel of Cultural Transmission left by the sum of past moments in the same area. Although several "time slices" are mixed up, it is not possible to isolate them (which is why we use "paths" in plural) unless we have detailed radiocarbon dates or some other chronological control. The difference between pattern (CTAPs) and processes arises from two sources: Cultural Transmission and taphonomic processes. Let us review them.

## 4.2 *Cultural Transmission Processes in Rock Art*

As a consequence of the bias forces implied in Cultural Transmission processes Henrich et al. (2008) has pointed out that, "ideas are not transmitted intact from one brain to another. Instead, the mental representations in one brain generate observable behavior, a 'public representation' in Sperber's terminology. Someone else then observes this public representation, and then (somehow) infers the underlying mental representation necessary to generate a similar public representation. The problem is that there is no guarantee that the mental representation in the second brain is the same as it is in the first. (…) Moreover, inferential processes often systematically transform mental representations, so that unlike genetic transmission, Cultural Transmission is highly biased toward particular representations. Following Sperber (1996), we call the representations favored by processes of psychological inference (including storage and retrieval) 'cognitive attractors'" (Henrich et al. 2008: 121). As a part of these processes, inferential transformation accounts for why some representations are favored over others. Instead, selective attention (see Chabris and Simons 2010) accounts for why individuals pay particular attention to some individuals or events and not to others. In addition, error occurs during each copying or replication. The presence of error is based on inherent constraints not only on human perception but also on motor abilities. For instance, the percentage of error in lithic manufacture has been set at around 3–5 % (e.g., Eerkens 2000).

As culturally transmitted behaviors, recurrent rock art motifs are probably constituted as cognitive attractors, and the process that constitutes them as such has to do with copying errors, inferential transformation, and selective attention. Also intentional factors, such as element additions or subtractions on a motif, could produce totally new motifs. As studies dealing with pictorial stimuli have shown, "there is a general reduction in the length or complexity of the material, that much of the detail is lost, and only the overall impression of the material is preserved" (Mesoudi 2008: 93) Still, none of these studies have dealt with rock art motifs, so it still remains to be studied how content bias, for instance, affects its replication (Scheinsohn and Caridi in prep.) Beside the need to model how the transmission process functions in a systemic context, there are other issues that require attention in terms of the archaeological context.

### 4.3 Taphonomic Processes in Rock Art

The passage of time between the inscription of a motif on a rock and its observation (days, years, or centuries after) introduces noise in two ways: (a) Attrition or losses of motifs: weathering provoked by wind, precipitations, and erosion could affect the rock art support, allowing for the loss of motifs or the destruction of painted areas, (b) accretion or addition of motifs: rock art motifs could be the result of many painting or engraving events at the very same site. What we find, after many years, could be the result of accretion (continuous or discrete episodes) or a single event. This can be identified only by the existence of superposition among motifs in a certain site, but its absence does not mean that there is no accretion but simply that accretion is not recognizable.

Rock art stores information, and, as a signal, it has the advantage of great longevity. But the lack of chronological control introduced by accretion and/or attrition processes at the site scale intrudes the tracking of lineages in a direct way. To do this, we have to add the process that leads to CTAPs. Mutual Information allows us to solve this problem. But before proceeding to methodological issues, we will first contextualize our case study.

## 5 The Archeological Problem: Late Holocene Rock Art in NW Patagonia

### 5.1 Patagonia and Its Native Peoples

In the Southern Hemisphere, Patagonia is the only landmass projecting southward from 46° latitude on. Its shape determines an oceanic influence which creates a lack of tundra and permafrost in spite of its high latitude (Morello 1984). Due to the rain shadow effect provoked by the cordillera, or Andean mountain range, Patagonia is divided into two contrasting environments: a forested and rugged area present on both sides of the Andes and a stepped area, which covers most of its surface. Nowadays the Andes also divide Patagonia politically, between Argentina and Chile (see Fig. 1).

Patagonia was peopled around 11,000 BP (Borrero and Franco 1997). The steppe has been occupied continuously since then. The role of forested environments for hunter-gatherers is widely debated (see Bellelli et al. 2003), but in any case, the Andean forest shows occupational discontinuities until approximately 3000 BP, when the archaeological signal strengthens and becomes continuous. Europeans arrived in Patagonia during the sixteenth century. Spanish settlers occupied single spots in the Western slope of the cordillera (Chile). Some of them did not thrive and returned home, abandoning their livestock, which became *cimarrón* (wild). That was the cattle that native Patagonians initially adopted. What North American anthropologists called the "horse complex" had its expression in South America with the introduction of habits related to the horse (Scheinsohn 2003). Ethnographically, the Eastern slope of the cordillera was inhabited by Tehuelches (Casamiquela 1965; Escalada 1949), hunter-gatherers who in the nineteenth century engaged in pastoral-

**Fig. 3** Rock art motifs from NW Patagonia: A Motif 8 - Character state 5 (see Appendix 2) Double opposed regular fret; B Motif 19 Character state 1 (see Appendix 2) Sun; C Motif 24 - Character state 5 (see Appendix 2) Linked rhombuses; D Motif 1 - Character state 2 (see Appendix 2) Aligned strokes

ist living related with European cattle. On the western side of the cordillera lived the Mapuches, horticulturists who engaged in weaving and metalworking and controlled horse trading. Both were defeated and almost exterminated by the Argentinean and Chilean armies, and their territories incorporated to the Modern State Nations at the end of the nineteenth century. The survivors were turned into rural workers.

## *5.2 Late Holocene Rock Art in Patagonia*

After this brief review, we may say that, before the European colonization, the Late Holocene period in Patagonia is supposed to involve a relative increase in population density and a demographic expansion (see Barrientos and Perez 2004 among others). This process should correlate with restricted residential mobility and wide exchange networks, as indicated by obsidian exchange and other items traded with non-hunter neighbors (see Scheinsohn 2003).

Late Holocene rock art sites are characterized by a single style called "Estilo de grecas" (or Fret style; see Menghin 1957), also termed Complex Geometrical Abstract Trend (CGAT; see Gradin 1999). This geometric style is identified by broken lines forming complex stepped-crenellated patterns (called *grecas* or frets Fig. 3a), resembling labyrinths or bounding motifs. They are accompanied by zigzags (13, Appendix 2), circles (5, Appendix 2), rhombuses (24, Appendix 2), crosses (21, Appendix 2), tridents (46, Appendix 2), squares (22, Appendix 2), and other polygons (see Appendix 2 and Fig. 3b–d). The style has a wide distribution in Patagonia (from 36° to 47° south latitude and from the Atlantic coast to the cordillera).

Many interpretations have been postulated to explain the processes that led to this unique rock art style, the CGAT. At their core, they can be reduced to two opposing models (Scheinsohn 2011): (1) Broad-scale model: where the CGAT style reflects a wide interaction network at a macroregional level, with no internal differentiation, and (2) territorial model: Under this view, the CGAT reflects territorial circumscription and ethnic differentiation in the context of increases in population density, territorial sizes, or home range reduction. The hypothesis that the observed rock art differences could be interpreted in terms of territorial marking was not supported by an endemicity analysis performed in previous work (Scheinsohn et al. 2009; Scheinsohn 2011).

Our picture of Northwestern Patagonia in late Holocene times is one of hunter-gatherers trying to maintain links between different places. Most researchers agree that during that period, a steppe-based population had been incorporating forested environments, a process in which rock art would have been part of the colonizing social repertoire, and shared graphics would be present across thousands of kilometers (a similar case as the one posited for the initial settlement of the arid zone of Australia, McDonald and Veth 2011). The establishment and maintenance of regional social ties has been recognized as an important part of hunter-gatherer adaptations to uncertain environments, in terms of creating a "safety net" of contacts and relations that can be critical to survival (Whallon 2006).

## 5.3 *The Question of Meaning in Patagonian Rock Art*

Although many researchers adhere to the perspective that rock art was made during ritual activities (Whitley 2005, 2011), we think that it is problematic to assume this by default in the Patagonian case. Even though ritual interpretations have been proposed for some sites (Carden 2009), others have been effectively associated to domestic activities (Aschero 1996) indicating mixed situations. But we can also assume that rock art practices were interrupted at some point after the European arrival in Patagonia, because they were never witnessed by the authors of historical sources. In addition, written documents report that the Tehuelche said that the rock art had been made by their ancestors in the very remote past (or even by mythical beings; see references about Elengassen in Moreno 1997 and Claraz 2008). Some sources suggest that the Tehuelche used to avoid rock art sites, because staying there provoked madness (Millán de Palavecino 1963: 429) or bad luck (Castro 2010: 93). This indicates a lack of continuity between the historical Tehuelches and the rock art, hindering our possibility to interpret the meaning of the motifs. Although we acknowledge the meaningfulness of rock art motifs, we have decided to set the question of meaning aside and treat our NW Patagonian rock art motif database without any reference to meaning, since it is not relevant for our purposes.

# 6 Methods and Materials

## *6.1 Mutual Information Network (MIN)*

In this work we will define the MIN for a rock art motif database from Northwestern Patagonia. Using this MIN, we will track the CTAPs in the study area. In order to do so, we have selected a set of highly correlating pairs of motifs which we define as a cluster in the MIN. For each cluster we will analyze the sites where those motifs are present and we will establish a link between two sites when they share two or more state of characters. When those links are mapped into a geographical region, we obtain a Site Network (SN) associated to that cluster (set of motifs). Based on geographical data, the SNs could result in (1) different spatial territories (if they do not overlap spatially) or (2) overlapping spaces.

# 7 Database

We have analyzed 49 rock art archaeological sites from northwestern Patagonia, located in the steppe, the forest, and their ecotone in a study area located between the 40° 10′ and 45° 50′ parallels (600 lineal km from North to South; see Fig. 1). This is a small area within the whole recorded distribution of the CGAT style. We have organized the studied sites in nine regions which are detailed in Table 1. We have considered only the painted motifs. Engravings were left aside in order to avoid noise due to issues related to manufacture techniques. This decision does not seem to affect the sample representativeness. The most recent revision of rock art techniques in Patagonia (Fiore 2006) considers that from the whole of the sample (including all sites since human peopling), only 22 % are engraved and very few (7 %) record combinations of paintings and engravings. Even if we restrict ourselves to the CGAT style, it is mostly composed by painted motifs (Fiore 2006: 46).

All the rock art sites included in our database are rock shelters.[1] The lack of organic content in the paintings impedes direct dating. So we have assumed that the presence of the CGAT style has a chronological content related to the Late Holocene period, as suggested previously by various researchers (Belardi 2004; Podestá et al. 2008 among others). Moreover, for Fiore (2006), the preservation of the CGAT style attests to its lack of antiquity.

[1] Parts of this database have been previously used in Scheinsohn et al. (2009, 2015).

Every motif present in each site was tabulated (see Appendix 2 and Scheinsohn et al. 2009). Departing for a "lumper" classification (see Scheinsohn et al. 2015 for details) of 59 motifs, each motif type was considered a "character" that may or may not be present in a given site and could have different states of character. For example, the "circle" motif type included all those figures described by a circle, with 13 distinguishable states, including "filled circle," "empty circle," "circle with a point," or their spatial disposition (in groups, aligned, etc.; see Appendix 2, character 5). Coding was carried out by defining different states on the basis of the morphological aspects of the figures found at each site. In order to minimize the degree of subjectivity in the process of assigning a figure to a motif type and the codification of each state of character, each figure underwent intersubjective testing. Three operators, separately, identified a figure, either from a narrated description or illustration, with a motif type. Afterward, it was verified if the figure was assigned to the same motif type by each operator. In case of disagreement (i.e., that the same figure was assigned to different motif types by different operators), the figure was identified by the majority's opinion (see details in Scheinsohn et al. 2015).

It should be noticed that although we have analyzed all the published sites for this study area, the whole universe of sites is unknown since it is a reasonable guess that many rock art sites are still undiscovered, especially in the forested part of the area, due to archaeological visibility problems (see Scheinsohn 2011; Scheinsohn and Matteucci 2013).

Also, as is characteristic of decorative patterns, in this dataset we have a quantity of motifs which in their majority are absent from most sites (Scheinsohn et al. 2009, 2015; see also Shennan and Bentley 2008 for ceramic decoration) and whose frequencies are distributed "with a large number of variants occurring only in small numbers but a small number being copied frequently and thus occurring a large number of times" (Shennan and Bentley 2008: 170). Since decorative patterns are unconstrained, variability is high. This results in a database with many "zero" data which posits an analytical challenge and leads us to a specific data treatment.

## 7.1 Mutual Information Network (MIN) Construction

In order to avoid the noise introduced by the "zero" data, we considered a dataset of 43 motifs (those of the original 59 motifs of the dataset used in Scheinsohn et al. 2015 which are present in three or more sites) of 49 assessed sites. We defined a variable associated with each of the motifs which received value 1 when the motif was present (no matter the state of character of the motif) and value 0 when it was absent from a particular site. For example, $X_1$ variable contains the information of motif 1; thus $X_1 = (1,1,1,0,0,0,\ldots 1,0,0)$ (this motif is present in sites 1, 2, 3 and absent in sites 4, 5, 6, etc.), $X_2$ variable contains information of motif 2 and so on. We performed the calculus of Mutual Information for each pair of motifs, and we selected the correlated pairs of motifs that had greater values of mutual information. A Threshold value for Mutual Information called $u$ and we fix it (set it, established it) in 0.093, which left 2.5 % of cases showing a value greater than $u$ (see details in Appendix 3). In Table 2

**Table 2** Pairs of motifs selected to establish links on the MIN

| Motif *X* number | Motif *Y* number | Mutual Information | Presence of *X* | Presence of *Y* | Sense of correlation |
|---|---|---|---|---|---|
| 50 | 15 | 0.093 | 3 | 5 | + |
| 33 | 26 | 0.093 | 5 | 3 | + |
| 56 | 48 | 0.093 | 3 | 5 | + |
| 36 | 8 | 0.094 | 3 | 31 | – |
| 21 | 4 | 0.096 | 24 | 11 | + |
| 38 | 11 | 0.096 | 13 | 10 | + |
| 5 | 2 | 0.097 | 32 | 7 | + |
| 6 | 5 | 0.097 | 7 | 32 | + |
| 25 | 14 | 0.099 | 3 | 17 | + |
| 22 | 11 | 0.104 | 9 | 10 | + |
| 33 | 3 | 0.105 | 5 | 13 | + |
| 27 | 25 | 0.105 | 16 | 3 | + |
| 27 | 10 | 0.112 | 16 | 6 | + |
| 53 | 24 | 0.114 | 14 | 10 | – |
| 23 | 2 | 0.124 | 18 | 7 | + |
| 36 | 3 | 0.126 | 3 | 13 | + |
| 50 | 38 | 0.126 | 3 | 13 | + |
| 8 | 50 | 0.134 | 12 | 3 | + |
| 56 | 54 | 0.134 | 3 | 12 | + |
| 12 | 1 | 0.138 | 9 | 31 | + |
| 25 | 20 | 0.143 | 3 | 11 | + |
| 23 | 21 | 0.145 | 18 | 24 | + |
| 35 | 22 | 0.156 | 5 | 9 | + |
| 24 | 13 | 0.284 | 10 | 22 | + |

In columns 1 and 2, the two motifs involved in the pair, which were generically called *X* and *Y* Motif number is referred to Appendix 2; column 3 is the Mutual Information for that pair $I(X, Y)$; and in column 4, the sense of the correlation is expressed with (+) when positively correlated and with (–) when negatively correlated

we show the 24 pairs of selected motifs, the presence of each of the motifs of the pairs, and the value of the Mutual Information obtained.

We used these 24 pairs in order to construct the MIN and define clusters in it. The threshold defines the links we use for defining the MIN. If we decreased or increased the threshold, we would define more or less links. This would modify the cluster structure on the network.

In our case, the Mutual Information could be a positive (if *X* is present, *Y* is present) or negative correlation (if *X* is present, *Y* is absent), given the variable has two possible states: presence or absence of the motif.

As we explained previously, we defined an associated Site Network (SN) for each cluster. We established a link between two nodes (sites) if they shared two or more character states of the motifs in the cluster, where the characters could be part of the same or of different motifs. Next, we analyze the structure of each resulting SN in terms of the degree of distribution and clustering coefficient, which measures the transitivity of the network by quantifying how frequently two neighbors of a node are each other's neighbor. We finally analyze the geographical location of each network to evaluate if there is overlapping between SNs, using the RgoogleMaps package of R project (R Development Core Team 2008; Loecher 2012). To simplify, we grouped sites which are closer than 5 km as one point in the map.

# 8 Results

Figure 4 shows the MIN obtained for the threshold $u$. As it can be seen (also in Table 2), only two links are negatively correlated; the rest of the links represent positively correlated pairs of motifs. We can observe six clusters on the MIN (the

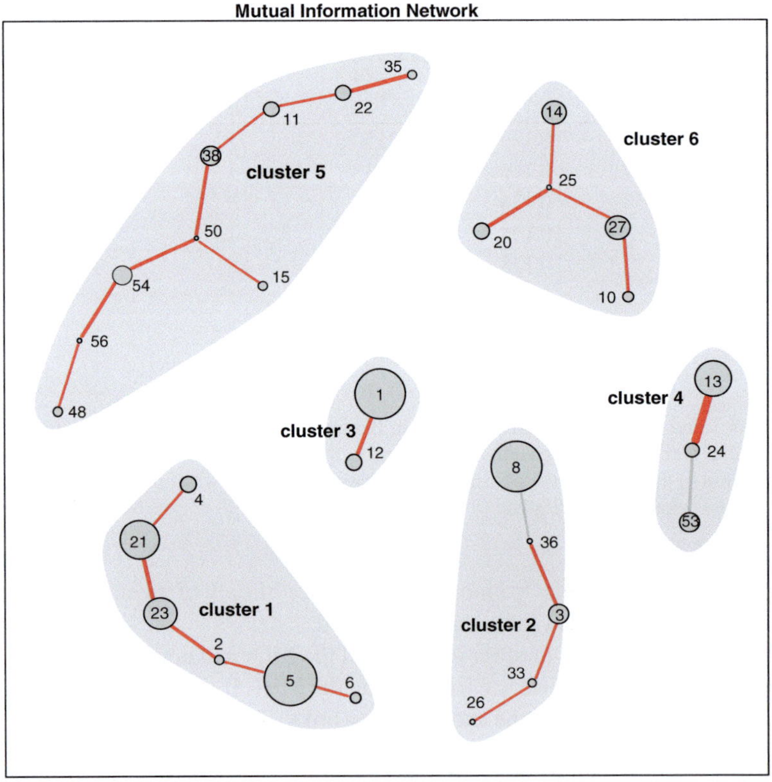


**Fig. 4** Mutual Information Network of motifs plotted using the igraph package for the R project (R Development Core Team 2008; Csardi and Nepusz 2006). Nodes represent motifs, and links represent the most correlated set of motifs in terms of presence/absence (without the detail of the states of the character of the motif). The size of the nodes is proportional to the presence of the motif in the sites of the studied area. The width of the links is proportional to the value of the Mutual Information between motifs. Bold links (*red* in the digital version) represent positively correlated motifs, while soft-colored links (*grey* in the printed version) represent the two cases of negatively correlated motifs

clusters are shaded in Fig. 4). Each cluster represents a set of independent motifs. Note that clusters are separated because we are not considering values of Mutual Information below *u*.

Below we analyze each cluster of the MIN and its associated SN:

- Cluster 1 links motifs 4, 21, 23, 2, 5, and 6 (see Appendix 2) in a lineal way. In Fig. 5 we see the Site Network associated to cluster 1, and in Fig. 6 we can see the same network in its geographical location. There are 30 sites which compose the SN. The connectivity results in a mean degree of 8.27, which means that each node (site) has, on average, approximately 8 neighbors. Nevertheless, the most frequent degree value is 5. Additionally, over half of the nodes have a degree of less than or equal to 5, and there are 7 nodes highly connected with more than 13 neighbors each. The most connected site is Cerro Pintado (CP) linked with all the other sites except two. Other sites with high degree values are Quila Quina 1 (QQ1, 19 neighbors), Risco de Azócar 2 (RA2, 18 neighbors), Puerto Tranquilo

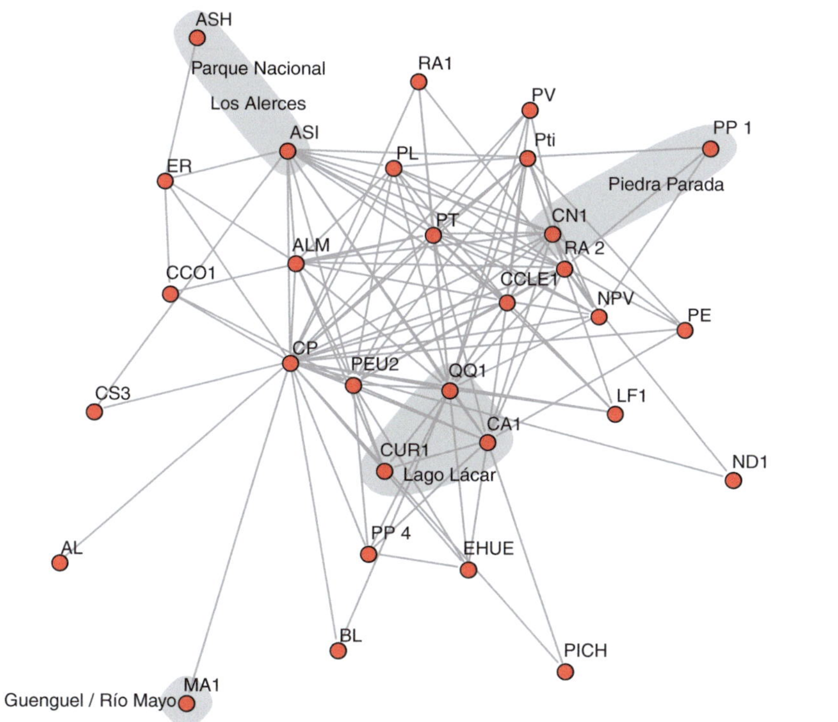


**Fig. 5** Cluster 1 SN plotted using (R Development Core Team 2008) igraph. Nodes represent archaeological sites; links are established when two sites share at least two character states of motifs from cluster 1. There are 12 sites isolated that were removed from the figure because they did not fit this last requirement, and only those nodes which had at least one link on the network were represented. Sites located less than 5 km from each other and Manantial 1 (MA1) are shaded, and the name of the corresponding region has been added

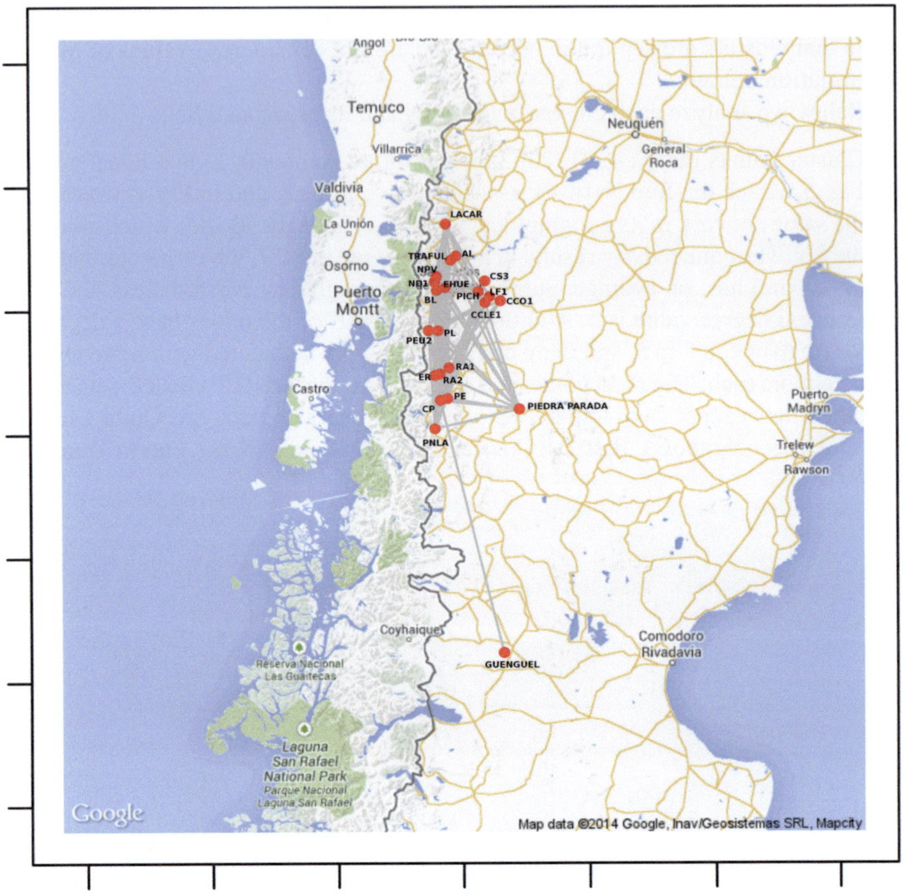


**Fig. 6** Cluster 1 SN with nodes geographically located, using the RgoogleMaps package of R project (R Development Core Team 2008; Loecher 2012). Nodes represent archaeological sites. In those cases in which two sites were less than 5 km apart, we joined and labeled them with the region's name, for example, *PP1* Piedra Parada 1, *PP4* Piedra Parada 4, and *CN1* Campo Nassif 1 are shown as Piedra Parada region

(PT, 16 neighbors), Peumayén 1 (PEU2, 15), Alero del Sendero de Interpretación (ASI), and Cueva Cuadro Leleque 1 (CCLE1, both with 14 neighbors). The clustering coefficient of this network is 0.56. A clustering of a random network with the same number of nodes and links but randomly allocated would be 0.28 approximately, which means that in this SN, the probability of two neighbors of a node being connected to each other is approximately two times the randomly allocated links. From a geographical standpoint, this network integrates all the regions of the study area but one (Lago Verde/Río Pico; see Fig. 1 and Table 1).

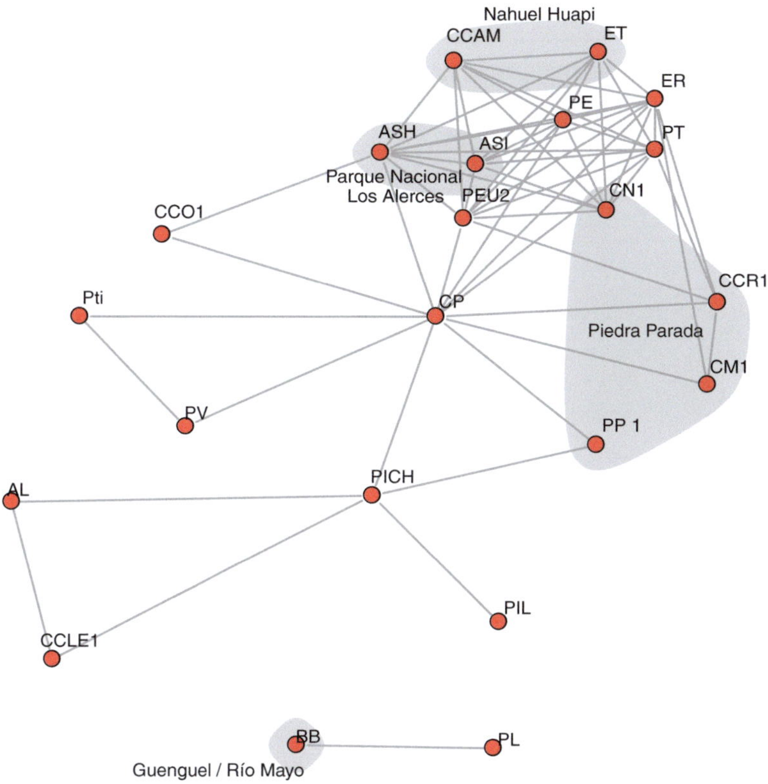


**Fig. 7** Cluster 2 SN. Nodes represent archaeological sites; links are established when two sites share at least two character states of motifs from cluster 2. There were 17 isolated sites that were removed from the figure because they did not fulfill this last requirement

- Cluster 2 is composed by the following motifs: 8 (fret or *greca;* see Fig. 3a), 36, 3, 33, and 26 in a chain. The outstanding fact is that, other than the fret (*grecas*), this network is composed by animal footprints. Also, the relationship between frets and human feet is negatively correlated: each time a fret is present, human feet are not (but notice that human feet are present only in three sites in this sample). In Fig. 7 we can see the cluster 2 SN, which include 22 sites. In this case we can see two clusters (one of two sites and the other with the rest). The sites with a maximum degree are Cerro Pintado (CP, 12 neighbors), El Radal (ER 11 neighbors), Peumayén 2, Puerto Tranquilo y Alero del Shamán (PEU2, PT, and ASH, respectively, 10 neighbors each) although the mean degree is 5.5. Cerro Pintado plays a fundamental role in ensuring the connectivity of the network. If we remove it, the network would split in three disconnected clusters. The clustering coefficient of this network is 0.74; the expected clustering coefficient of a

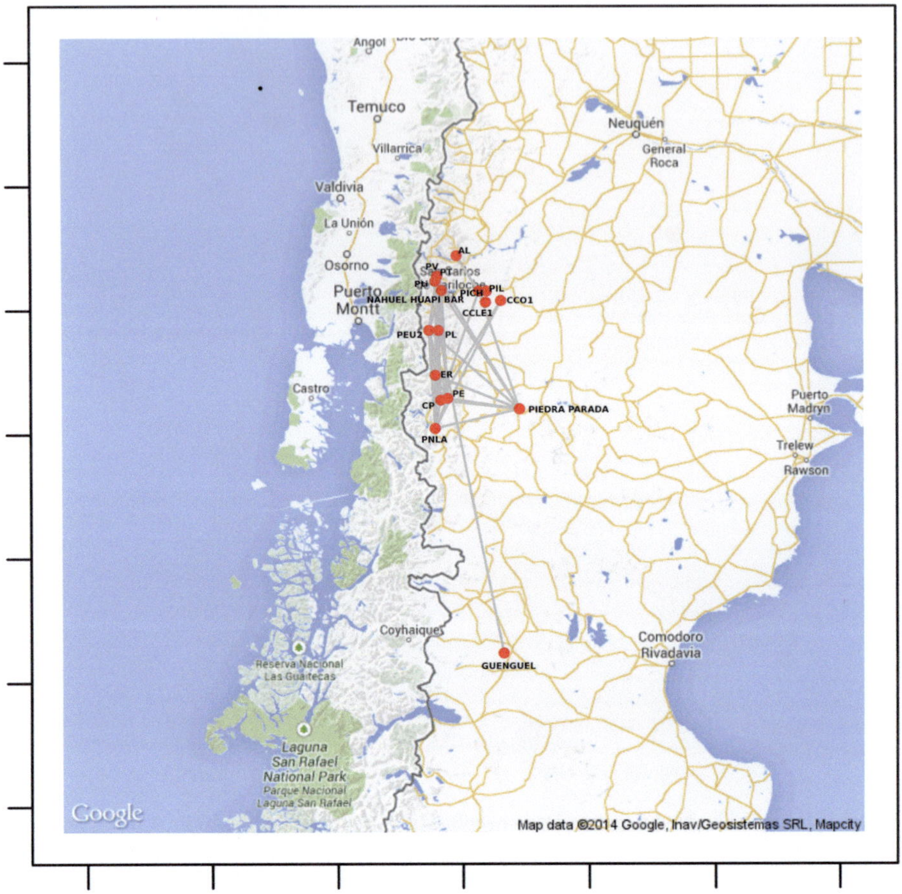


**Fig. 8** Cluster 2 SN with nodes geographically located (R Development Core Team 2008; Loecher 2012)

random network with the same number of nodes and links but randomly allocated would be 0.25 approximately. In this case the transitivity of the SN is more important than in the previous one. The interconnectedness of the superior sector of the figure is also surprising, which includes Alero del Shaman (ASH), Cerro Campanario (CCAM), El Trébol (ET), El Radal (ER), Puerto Tranquilo (PT), Campo Nassif (CN1), Peumayén 2 (PEU2), Alero del Sendero de Interpretación (ASI), and Peñasco (PE). This set of sites correspond to the northern region of the study area and are aligned from north to south – with the exception of Campo Nassif 1 (CN1) located at the east; see network on the map in Fig. 8. These sites

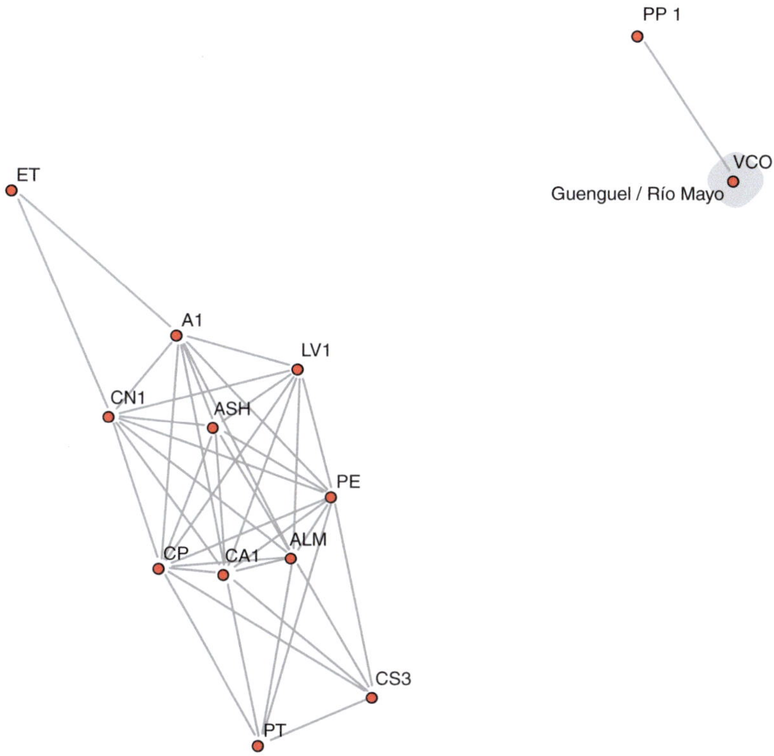


**Fig. 9** Cluster 3 SN. Nodes represent archaeological sites; links are established when two sites share at least two character states of motifs from cluster 3. There were 18 isolated sites that were removed from the figure because they did not fulfill this last requirement

share the same pairs of character states for the fret motif. Cluster 2 SN presents all the regions that constitute the study area except for Lago Verde/ Río Pico in the south and Lago Lácar in the north. The Guenguel/Río Mayo region forms a separate cluster (with Bardas Blancas, BB) linked to the core of the geographical region by Paredón Lanfré (PL).

- Cluster 3 links only two motifs: 1 and 12 (see Appendix 2). Cluster 3 SN (see Fig. 9) links few sites highly interconnected in one case and two other sites connected by only one link (superior part of Fig. 9). Guenguel/Río Mayo (Viejo Corral, VCO) again remains attached by only one link, in this case to Piedra Parada 1 (PP1). The wide range of regions in this network is noteworthy (Fig. 10): all the regions that constitute the study area are represented. Although in the SN there are two disconnected structures, two nodes (Piedra Parada 1, PP1, and

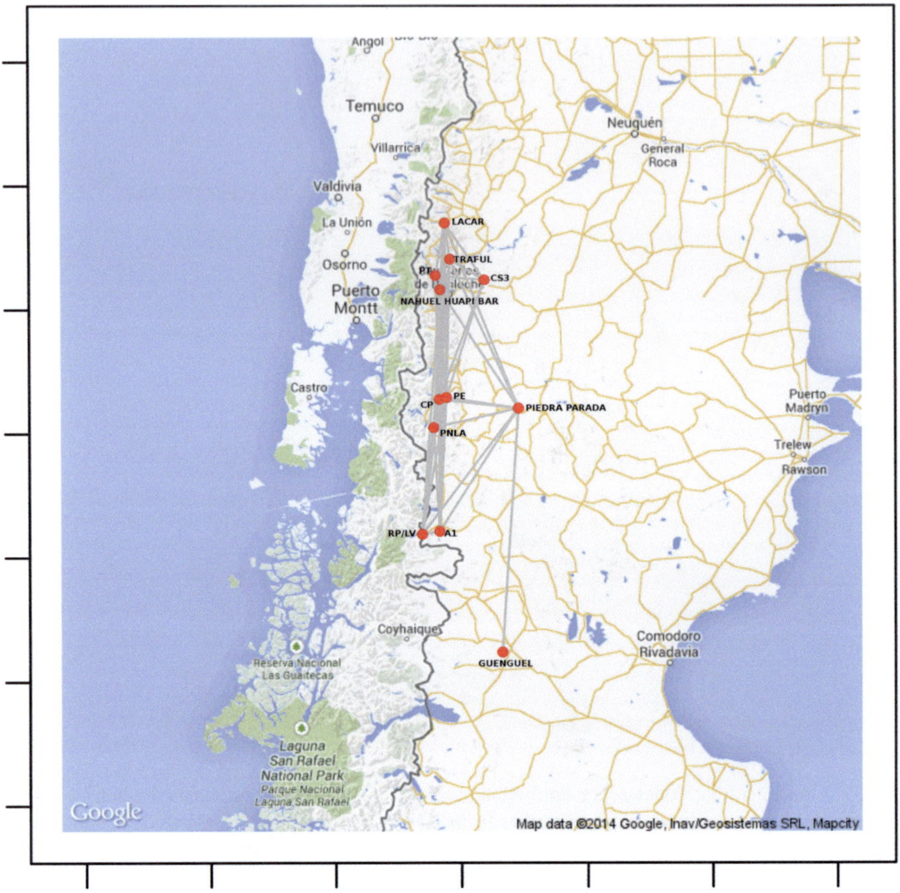


**Fig. 10** Cluster 3 SN with nodes geographically located (R Development Core Team 2008; Loecher 2012)

Campo Nassif 1, CN1) are less than 5 km from each other linking the Piedra Parada region with the rest. The clustering coefficient of this network (with a value of 0.83) is twice the value of a network with random assignment links. Cluster 3 SN is the only one which connects the whole area (9 regions).

- Cluster 4 links three motifs: 13, 24, and 53 (see Appendix 2). Motifs 53 and 24 are negatively correlated. Risco de Azócar 1 (RA1) is the most connected site (with 9

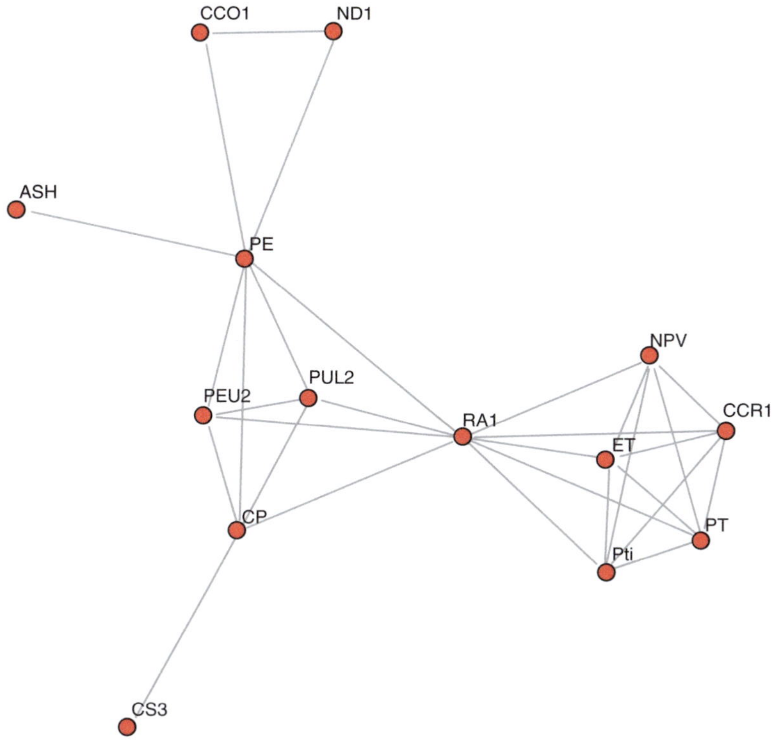


**Fig. 11** Cluster 4 SN. Nodes represent archaeological sites; links are established when two sites share at least two character states of motifs from cluster 4. There were 15 isolated sites isolated that were removed from the figure because they did not fulfill this last requirement

neighbors) and plays an important role in connecting two different structures in cluster 4 SN (see Fig. 11). The clustering coefficient of this network is 0.71 (a value of approximately 0.30 would be obtained if the same number of links was randomly allocated on the network). In Fig. 12 we can see the sites located on the map. From the nine regions which make up the study area, only five are represented on this network, leaving aside the extreme north (Lago Lácar and Traful) and the extreme south (Lago Verde/Río Pico and Guenguel/Río Mayo).

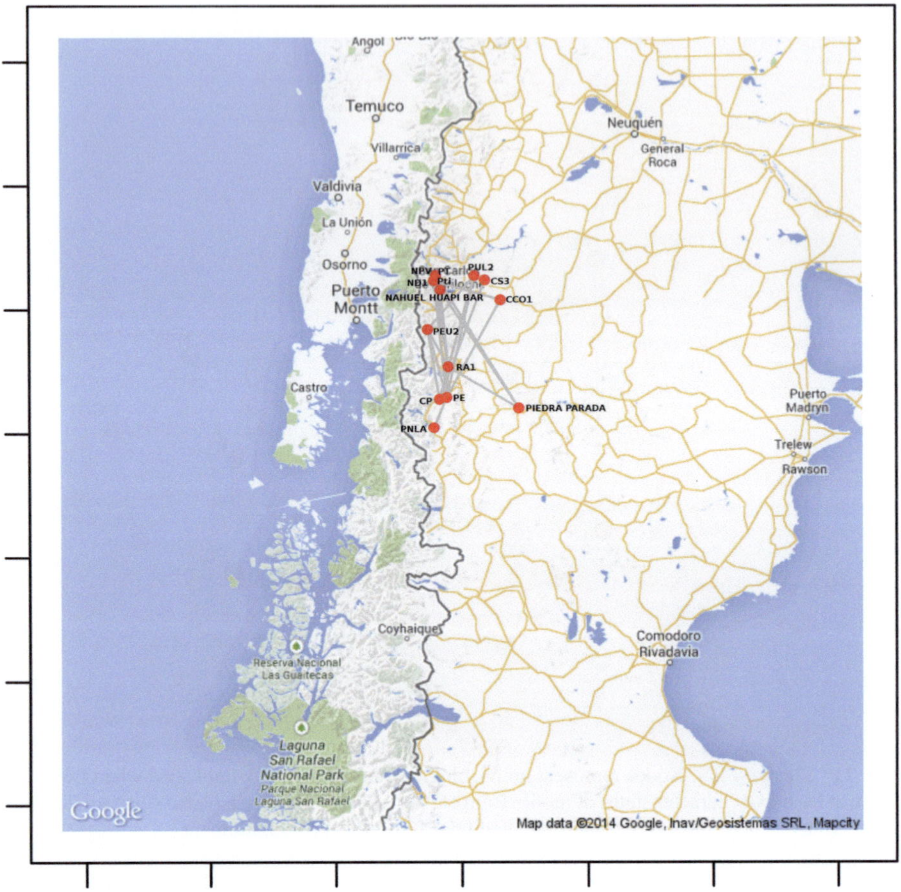


**Fig. 12** Cluster 4 SN with nodes geographically located (R Development Core Team 2008; Loecher 2012)

- Cluster 5 links the following motifs: 35, 22, 11, 38, 15, 50, 54, 56, and 48 (see Appendix 2). It is characterized by figurative (35, 38, 15, 56, and 48) and geometric motifs (22, 11, 50, 54). This cluster has the highest quantity of motifs but with few members each, and it is, with cluster 6, the only one that

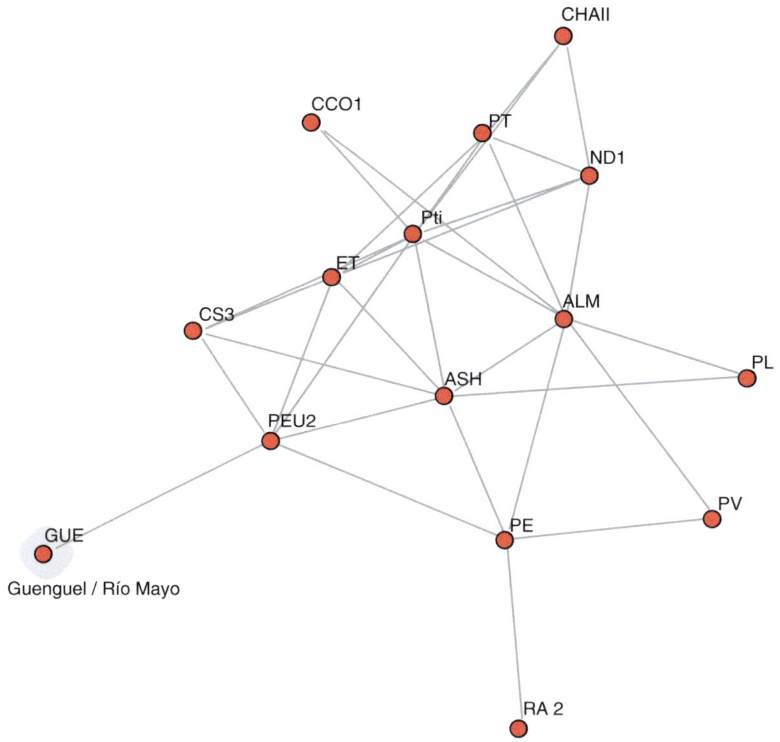


**Fig. 13** Cluster 5 SN. Nodes represent archaeological sites; links are established when two sites share at least two character states of motifs from cluster 5. There were 14 isolated sites isolated that were removed from the figure because they do not fulfill this last requirement

does not follow a chain pattern. Motif 50 actually functions as a hub in the cluster. In Fig. 13 we can see cluster 5 SN. Puerto Tigre (Pti) is the most connected node, but its absence would not separate further clusters. Figure 14 presents the sites in the map. There are six regions included. The excluded

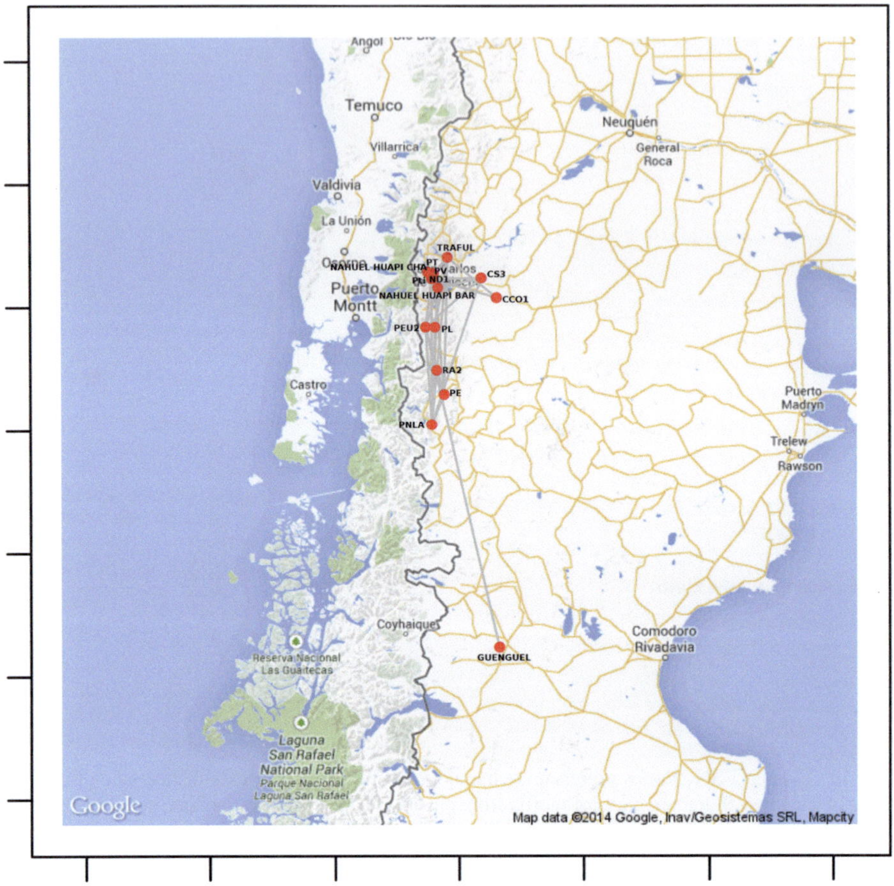


**Fig. 14** Cluster 5 SN with nodes geographically located (R Development Core Team 2008; Loecher 2012)

regions are Lago Lácar, Piedra Parada, and Lago Verde/Río Pico. As in the case of cluster 2, Guenguel/Río Mayo is linked to the same region (in this case, by a link with Peumayén 2, PEU2, and not Paredón Lanfré, PL but these sites are very close to each other).

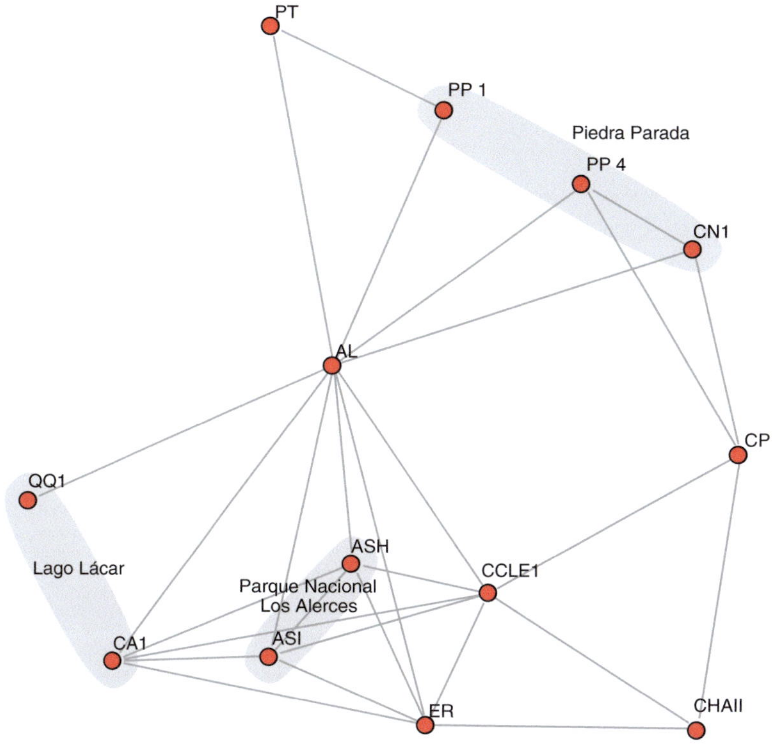


**Fig. 15** Cluster 6 SN. Nodes represent archaeological sites; links are established when two sites share at least two character states of motifs from cluster 6. There were 16 isolated sites that were removed from the figure because they do not fulfill this last requirement

- Finally cluster 6 is comprised by five motifs: 10, 27, 25, 20, and 14, disposed in a Y pattern with 25 functioning as a hub. In cluster 6 SN (Fig. 15), the Alero Lariviere (AL) site, located at the north, is the most connected node, and it forms the link between two groups that would otherwise be separated. The inferior part

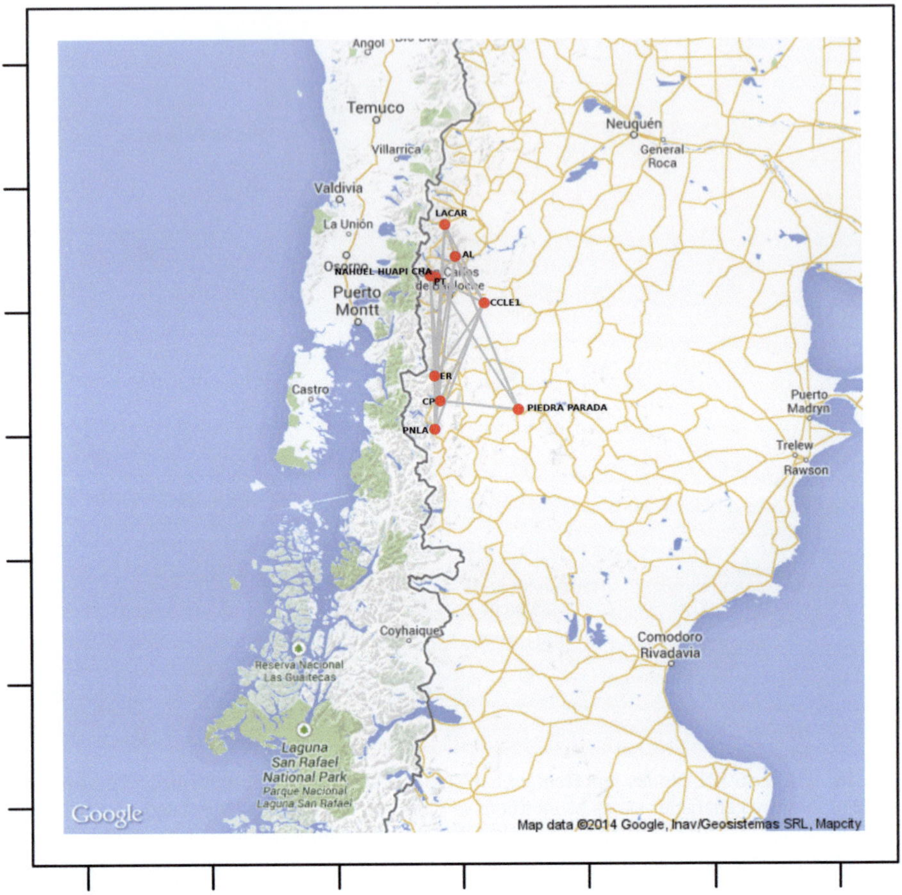


**Fig. 16** Cluster 6 SN with nodes geographically located (R Development Core Team 2008; Loecher 2012)

of the SN is highly interconnected. Figure 16 shows its wide geographical dispersion. In this cluster, the two southern regions (Lago Verde/Río Pico y Guenguel/Río Mayo) are not represented.

## 9 Discussion

The presented results hinge on the threshold for Mutual Information between motifs that was established in order to construct the MIN (therefore, the clusters of defined motifs) and are based on the requirements set for defining the SN (at least two character states had to be shared to establish a link between sites, a very strong condition).

Our results allowed us to detect a strong overlapping in all the SNs and in turn, the SNs connected widely dispersed sites (in the order of 600 km in some extreme cases). As a consequence, we could not separate territories.

We consider that each SN could be interpreted as a different CTAP (see Fig. 2, bottom). Since CTAPs are an archaeological signal of one or many CTPs (see Fig. 2), we believe that our MIN clusters allow us to differentiate CTAPs with temporal and spatial dimensions even if we cannot determine the duration of each chronological "slice" or their sequence (i.e., which one was first and which last). The links that are part of the path could be contemporaneous, or not. In that sense, we cannot trace actual phylogenetic relationships between them, only undirected networks.

The fact that most of the SNs are spatially overlapping for each MIN cluster, and that there are sites that appear connected in many CTAPs, allows for the redundancy of the flow of information. This pattern is stronger in the middle and northern parts of the study area. Redundancy and high connectivity among sites are compatible with accretion and, hence, with elapsed time. Given that some of the sites delimited in the study area are not in close geographical range, we can assume, in archaeological terms, that the middle and northern parts of the study area should be in an effective occupation/colonization phase (see Borrero 1994–1995), while the south part (Guenguel/Río Mayo) and the extreme north (Lago Lácar) should be in an exploration phase (less redundancy and connections with the rest; see Borrero 1994–1995).

The regions that are present in all the SN clusters are Pilcaniyeu, Nahuel Huapi Comarca Andina del Paralelo 42°, and Parque Nacional Los Alerces. In the first three cases, we can argue that in those regions, there are more sites recorded and probably this is due to a sample bias. But in Parque Nacional Los Alerces, there are only two sites. Then, preliminary, we argue that those four regions are the ones in which more time of human occupation elapsed. The regions of Traful and Piedra

Parada could be included because they are absent in only one cluster (Traful does not appear in cluster 4 SN and Piedra Parada in cluster 5 SN). It is remarkable that cluster 6 SN includes only northern sites (from Parque Nacional Los Alerces to the north). Then we can separate a nuclear area (Pilcaniyeu, Nahuel Huapi Comarca Andina del Paralelo 42°, Parque Nacional Los Alerces, Traful, and Piedra Parada) and three other regions (Lago Lácar, Guenguel/Río Mayo, and Lago Verde/Río Pico) that are either in course of incorporation (exploration/colonization phases *sensu* Borrero 1994–1995) or integrated to other CTAPs with a different nuclear area (i.e., North of Santa Cruz Province). This last possibility should be explored in the future.

Also, the importance of the nodes is variable. In the case of the first three clusters, Cerro Pintado (CP), which is located in the middle of the study area (Comarca Andina del Paralelo 42° region), appears as a hub (cluster 1, 2, and 3 SNs, with the same degree of other three sites). Its importance could be related to the fact that it is the biggest sample (see Table 1). But, in addition, in cluster 1 SN, its absence would disconnect the Guenguel/Río Mayo region, while in cluster 2 SN, its absence would imply the disconnection of Pilcaniyeu and Traful from the more interconnected structure. In cluster 4 SN, Risco de Azócar 1 (RA1) has the higher degree, although located in the same region of Cerro Pintado (CP), and in this case, the loss of this site would disconnect the network into two structures (one with Piedra Parada and Nahuel Huapi and the other one with Parque Nacional Los Alerces, Pilcaniyeu, and Comarca Andina del Paralelo 42°). In cluster 5 SN, the most connected nodes are Alero Las Mellizas (ALM) and Puerto Tigre (Pti), but they do not account for the connectivity of the network. Finally, in cluster 6 SN, the most connected node is Alero Lariviere (AL), while the loss of this node would not disconnect the regions in the network. Hence, although the importance of CP as a node is high, there are other nodes that are also important in the same region.

## 10 Conclusions

Our analysis has allowed us to formalize the Mutual Information Network between rock art motifs in Northwestern Patagonia. Rock art sites offer a "flattened" temporal dimension, recreating the idea of a palimpsest which Binford (1981) applied to the archaeological record. In a rock art site, time may be compressed. Hence, we have established a MIN among rock art motifs and considered different Site Networks as CTAPs. Since SN clusters are not explained by a territorial model (given the spatial overlapping between them), we have considered a MIN cluster's motifs as contemporaneous and/or representative of heritable continuity associated to CTAPs. Drawing on the rock art, we have determined CTAPs for a set of sites, by means of motif correlation. Probably these CTAPs will work also for the transmission of other kinds of information. Then, these CTAPs could be tested against other lines of evidence. In the future we will explore other conditions for the MIN (as changes in $u$) and SN requirements.

Furthermore, this analysis also allows us to relate rock art sites to Borrero's model for the population of Patagonia. We suggest that the strong connectivity between the middle and northern regions of the study area reveals a hypothetical nuclear region, well known and transited by the hunter-gatherer population who made the rock art. The few links with the extreme north and the extreme south regions allow us to maintain that those areas were in an exploration/colonization phase. Let us note that, as we have argued elsewhere (Scheinsohn et al. 2009, 2015), site proximity, in terms of geographical distance, did not ensure connections (in our case, motif sharing). So, another interpretation could be that the mix of high local connections and sparse regional connections is what sustained the information flux. Following White (2013), we can suggest that "the creation of a relatively sparse web of non-local connections is the most efficient way to engineer a significant improvement in the ease of information flow across a spatially-situated network (…) If the main purposes of maintaining non-local contacts is to facilitate 'over the horizon' information flow and secure access to assistance or resources in distant areas during times of stress, the cost of maintaining connections would be an incentive to have as few as necessary to serve the purpose of maintaining sufficient information flow" (White 2013: 20). The same has been proposed by Kauffman (1993), when investigating genetic networks. He suggested that genetic networks would have to be sparsely connected since densely connected networks seemed incapable of settling down into stable cycles. Nowadays, there is a good quantity of works from different fields (i.e., in biology Leclerc 2008, in sociology Granovetter 1973, etc.) which propose that a parsimonious network sparsely connected and not unnecessarily complex will be robust enough. Our results from rock art motifs fit this picture. Further research will be directed to exploring this issue and test whether these sites are best linked to other nuclear areas (North of the Santa Cruz province, for instance).

**Acknowledgements** We are grateful to Larissa Mendoza Straffon for inviting us to contribute to this volume. We would also like to thank Mercedes Salado Puerto (EAAF). This work would not have been possible without her. Sabrina Leonardt and Florencia Rizzo contributed to the database. Florencia Rizzo also helped with Figs. 1, 2, and 3. The Comarca Andina del Paralelo 42 Team (Cristina Bellelli, Mariana Carballido, Pablo Férnandez, Ana Forlano, and Mercedes Podestá) made their firsthand database available to us. Finally, we also thank the INAPL staff. Part of this research was funded by FONCYT PICT 2010 N° 1810 and CONICET.

# Appendix 1

Before defining Mutual Information, we need to introduce the Shannon Information and the Entropy functions. In his classic paper, Shannon (1948) defined entropy (*H*) as a measure of uncertainty of a random variable. In a communicational process, a given source emits messages that can be stored in an *X* variable. Departing from probability distribution of *X* values, entropy quantifies the level of "surprise" the receiver experiences upon receiving each message. If there is no surprise, there is no Information content (Mitchell 2009), because the message is fully predictable.

As an example, we present a rock art case. Let us introduce variable *X*, which represents a particular motif, which can take two possible values: 0 means the absence of the motif in a particular site and 1 its presence. Let us suppose that we have assessed 8 sites for this particular motif (see Table A1.1).

This motif is present in sites 1, 2, 3, and 4, and it is absent in the rest of the sites. From the data, we can compute the probability that *X* takes value 0 (which we will call $P(X=0)$) and the probability that *X* takes value 1 (which we will call $P(X=1)$). The probability of X taking a particular value is the frequency of this particular value with respect to the total number of observations. In this example, $P(X=1)=\frac{4}{8}=\frac{1}{2}$ (because in 4 of the 8 assessed sites the motif is present) and $P(X=0)=\frac{1}{2}$. Let us note that $P(X=1)+P(X=0)=1$ because *X* can take only two values in this example. The Shannon Information contained in the outcome value 1 of variable *X* is defined as

$$h(X=1)=-\log_2 P(X=1).$$

The entropy of the variable *X* is defined as the average of the Shannon Information contained in the possible outcomes, thus:

$$H(X)=P(X=1)h(X=1)+P(X=0)h(X=0).$$

The entropy of *X* for the example is $H(X)=1$, because

$$H(X=1)=-\frac{1}{2}\log_2\left(\frac{1}{2}\right)-\frac{1}{2}\log_2\left(\frac{1}{2}\right)=\frac{1}{2}+\frac{1}{2}=1$$

Let us note that:

- The entropy is always greater than or equal to zero. The particular case of entropy zero occurs when the variable *X* takes a particular value with probability 1 and

**Table A1.1** Example of variable *X* representing the absence (when *X* takes the value 0) and the presence (when *X* takes the value 1) of a particular motif in 8 different sites

| | Site 1 | Site 2 | Site 3 | Site 4 | Site 5 | Site 6 | Site 7 | Site 8 |
|---|---|---|---|---|---|---|---|---|
| Motif (*X* variable) | 1 | 1 | 1 | 1 | 0 | 0 | 0 | 0 |

**Table A1.2** Example of variables $X_1$, $X_2$, $X_3$ and $X_4$ representing the presence (when variable takes the value 0) and absence (when variable takes the value 1) of four motifs in each of th3 8 assessed sites

| | Site 1 | Site 2 | Site 3 | Site 4 | Site 5 | Site 6 | Site 7 | Site 8 | *H* (Entropy) |
|---|---|---|---|---|---|---|---|---|---|
| Motif 1 $X_1$ | 0 | 0 | 0 | 0 | 0 | 0 | 0 | 0 | 0 |
| Motif 2 $X_2$ | 1 | 1 | 1 | 1 | 1 | 1 | 1 | 1 | 0 |
| Motif 3 $X_3$ | 1 | 1 | 0 | 0 | 0 | 0 | 0 | 0 | 0.81 |
| Motif 4 $X_4$ | 0 | 0 | 1 | 1 | 1 | 1 | 1 | 1 | 0.81 |

**Table A1.3** Example of two variables *X* and *Y* representing the presence (when variable takes the value 0) and absence (when variable takes the value 1) of two motifs in 8 sites

| | Site1 | Site 2 | Site 3 | Site 4 | Site 5 | Site 6 | Site 7 | Site 8 |
|---|---|---|---|---|---|---|---|---|
| Motif 5 (*X*) | 1 | 1 | 0 | 1 | 1 | 0 | 0 | 0 |
| Motif 6 (*Y*) | 1 | 1 | 0 | 1 | 1 | 1 | 0 | 0 |

the rest of the values with probability 0. Then, the uncertainty of the variable is 0, because we are certain that *X* will take only one possible value.

- Entropy *H*(*X*) reaches its maximum value when the probability of the occurrence of the outcomes of *X* variable is uniform. For the particular case of two possible outcomes, *H*(*X*) is maximum when the probability of the two outcomes is the same, $P(X=1)=P(X=0)=\frac{1}{2}$, and it reaches the value $H(X)=1$. In this scenario, the uncertainty of the variable *X* is maximum.

Let us complicate our example a little bit further. We will continue with 4 motifs and 8 assessed sites (Table A1.2).

In the case of motif 1 and motif 2 (row 1 and row 2 of Table A1.2), the entropy takes the same value ($H(X_1)=H(X_2)=0$). This occurs because entropy is a function of the probabilities and not of the values of the outcomes. Entropy does not distinguish between two cases which are symmetric (if we change outcomes 1 for 0 and vice versa). The same happens with motifs 3 and 4 ($X_3$ and $X_4$) that reach the same entropy value, which results $H(X_3)=H(X_4)=0.81$.

Now let us compare two other motifs (*X* and *Y*) in the same 8 sites. Let us suppose that the observed values result in Table A1.3.

We are interested in detecting if there is any type of correlation between these two motifs. Does information about motif 5 give information about motif 6, or are they independent variables? Mutual Information helps to answer this question by quantifying the information gain that we obtain from one variable when we know the other variable and vice versa. The Mutual Information is defined as:

$$I(X,Y)=H(X)-H(X\mid Y)$$

where $H(X\mid Y)$ is the conditional Entropy of variable *X* given that we know the variable *Y* (Cover and Thomas 1991). As we just mentioned, Mutual Information measures the difference in the uncertainty of *X* variable when we know *Y* variable and vice versa. When variables *X* and *Y* are independent, then the fact of knowing *Y*

**Table A1.4** Example of two variables X and Y representing the presence (when variable takes the value 0) and absence (when variable takes the value 1) of two motifs in 8 assessed sites

| | Site 1 | Site 2 | Site 3 | Site 4 | Site 5 | Site 6 | Site 7 | Site 8 |
|---|---|---|---|---|---|---|---|---|
| Motif 7 ($X$) | 1 | 1 | 0 | 1 | 1 | 0 | 0 | 0 |
| Motif 8 ($Y$) | 0 | 0 | 1 | 0 | 0 | 0 | 1 | 1 |

does not reduce the uncertainty of *X*. Formally, $H(X|Y) = H(X)$, because the uncertainty of *X* is the same (regardless of whether *Y* is known or not). Then $I(X,Y) = H(X) - H(X|Y) = 0$, reflecting that the knowledge of one of the variables (*X* or *Y*) does not say anything about the other. In the other extreme case, the uncertainty of variable X is fully reduced when we know Y variable, $H(X|Y) = 0$ (if we know the value of *Y*, then the certainty of *X* variable is complete, because we are sure of the value which *X* takes), then the Mutual Information is maximum $I(X,Y) = H(X) - H(X|Y) = H(X)$.

Returning to the example of Table A1.3, we compute $I(X,Y) = 0.548$. Thus, *X* and *Y* are not independent. We can observe that every time motif 5 is present ($X=1$), then motif 6 is present too ($Y=1$) (notice that the inverse case is not met: in site 6 although motif 6 is present, motif 5 is absent).

Finally, it is important to remark that Mutual Information does not say anything about the sense of the information gain. Hence, in Table A1.4 we present other example which leads to the same value of Mutual Information as the previous example ( $I(X,Y) = 0.548$ ). But in this case, we can note that each time motif 7 is present ($X=1$), then motif 8 is absent ($Y=0$). Then, motif 7 gives us information about motif 8, but they are negatively correlated.

## Appendix 2

List of motifs and character states (Taken from Scheinsohn et al. 2009).

| Motif | Character state | Description | Design |
|---|---|---|---|
| 0 | 1 | Dot |  |
| | 2 | Aligned dots |  |
| | 3 | Grouped dots | 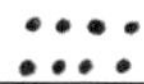 |
| 1 | 1 | Line stroke |  |
| | 2 | Aligned strokes |  |
| | 3 | Brush stroke |  |
| 2 | 1 | V |  |
| | 2 | Aligned Vs |  |
| 3 | 1 | Tridactyl |  |
| | 2 | Tridactyl in a geometric shape |  |
| | 3 | Aligned tridactyls |  |
| | 4 | United tridactyl |  |
| 4 | 1 | Z |  |
| | 2 | Aligned Zs |  |

| | | | |
|---|---|---|---|
| 5 | 1 | Empty circle |  |
| | 2 | Filled circle |  |
| | 3 | Circle with a point |  |
| | 4 | Concentric circle |  |
| | 5 | Circle with a cross |  |
| | 6 | Dotted circle |  |
| | 7 | Dotted concentric circles |  |
| | 8 | Aligned empty circles |  |
| | 9 | Aligned filled circles |  |
| | 10 | Aligned circles with a point |  |
| | 11 | Attached empty circles |  |
| | 12 | Circle with a stroke |  |
| | 13 | Concentric circles with a point | 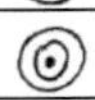 |
| 6 | 1 | Empty circles with a rod | 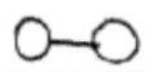 |
| | 2 | Concentric circles with a rod |  |
| | 3 | Empty circles with a zigzag rod |  |
| | 4 | Empty circles with attached elements in a rod | 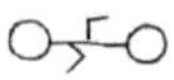 |
| | 5 | Concentric circles with attached elements in a rod |  |
| | 6 | Many Concentric circles with a rod | 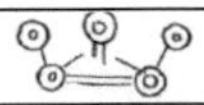 |
| 7 | 1 | Irregular closed figure | 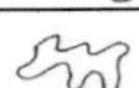 |
| | 2 | Grouped irregular closed figures | 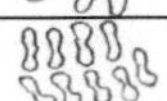 |
| 8 | 1 | Open irregular fret | 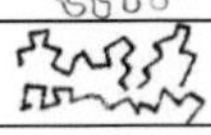 |
| | 2 | Open regular fret |  |
| | 3 | Double regular fret | 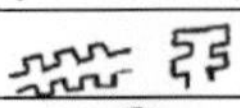 |
| | 4 | Closed filled fret |  |
| | 5 | Double opposed regular fret |  |
| | 6 | Closed empty fret | 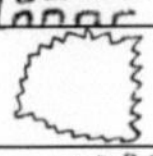 |
| | 7 | Closed empty fret with an inner figure | 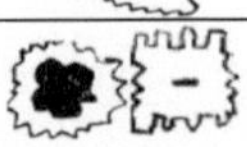 |

| | | | |
|---|---|---|---|
| 9 | 1 | Hollow ladder |  |
| | 2 | Staggered semicircle | 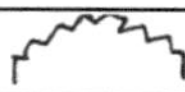 |
| 10 | 1 | Parallel staggered truncated pyramid-like lines |  |
| | 2 | Staggered truncated pyramid-like line |  |
| 11 | 1 | Comb-like figure | 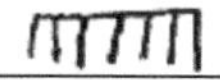 |
| | 2 | Double comb-like |  |
| 12 | 1 | Reticulated figure | 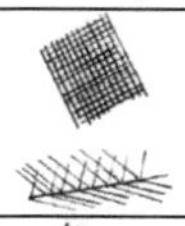 |
| | 2 | Reticulated rhombus | 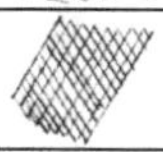 |
| 13 | 1 | Zigzag |  |
| | 2 | Aligned zigzag | 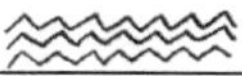 |
| | 3 | Zigzag strokes |  |
| 14 | 1 | Open arc |  |
| | 2 | Aligned open arcs |  |
| 15 | 1 | Undulated "tree" | 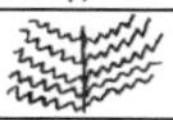 |
| | 2 | Totemic post |  |
| | 3 | Straight "tree" |  |
| 16 | 1 | Zigzag circle |  |
| | 2 | Zigzag circle with inner "sun" |  |
| | 3 | Irregular zigzag figure |  |
| 17 | 1 | Flying birds |  |
| 18 | 1 | drumhead |  |

| | | | |
|---|---|---|---|
| 19 | 1 | Sun |  |
| | 2 | Circled sun | 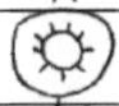 |
| | 3 | Sun with inner point |  |
| | 4 | Concentric sun |  |
| | 5 | Filled sun |  |
| | 6 | Partial sun |  |
| 20 | 1 | Oval figure | 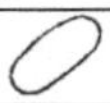 |
| | 2 | Oval with inner point | 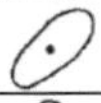 |
| | 3 | Oval with inner strokes |  |
| 21 | 1 | Simple cross |  |
| | 2 | Hollow cross |  |
| | 3 | Filled cross |  |
| | 4 | Concentric crosses |  |
| | 5 | Greek hollow cross | 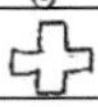 |
| | 6 | Trefoil hollow cross | 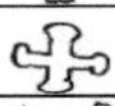 |
| | 7 | Staggered hollow cross |  |
| | 8 | Greek cross with inner filled cross |  |
| | 9 | Fretted inner cross |  |
| | 10 | Aligned simple crosses |  |
| | 11 | Aligned filled crosses |  |
| | 12 | Aligned concentric crosses |  |
| | 13 | Aligned fretted inner crosses |  |
| | 14 | Linked aligned simple crosses |  |
| | 15 | Aligned Greek cross with inner filled crosses |  |
| | 16 | Cross with ovals at their ends | 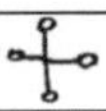 |
| | 17 | Others | |

| | | | |
|---|---|---|---|
| 22 | 1 | Square |  |
| | 2 | Subdivided square | 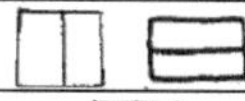 |
| | 3 | Square with two subdivisions | 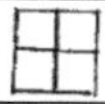 |
| | 4 | Concentric squares |  |
| | 5 | Square with inner strokes |  |
| | 6 | Filled square |  |
| 23 | 1 | Rect angle |  |
| | 2 | Subdivided rectangle |  |
| | 3 | Rectangle with two subdivisions | 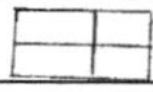 |
| | 4 | Concentric rectangle |  |
| | 5 | Rectangle with inner strokes |  |
| | 6 | Filled rectangle |  |
| | 7 | Rectangle with inner figure | 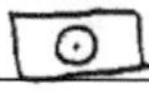 |
| | 8 | Rectangle with inner zigzag | 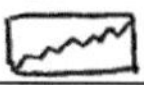 |
| 24 | 1 | Rhombus |  |
| | 2 | Rhombus with inner strokes |  |
| | 3 | Concentric rhombuses |  |
| | 4 | Aligned rhombuses |  |
| | 5 | Linked rhombuses |  |
| | 6 | Rod linked rhombuses |  |
| | 7 | Rhombus with attached element |  |
| | 8 | Staggered rhombus with inner circle | 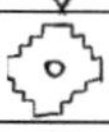 |
| 25 | 1 | Polygon | 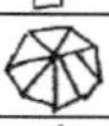 |
| | 2 | Irregular polygon |  |
| | 3 | Polygon with inner rectangle |  |
| 26 | 1 | Guanaco footprint |  |

| | | | |
|---|---|---|---|
| 27 | 1 | Hollow clepsydra |  |
| | 2 | Clepsydra with inner strokes |  |
| | 3 | Framed Clepsydra |  |
| | 4 | Staggered clepsydra |  |
| | 5 | Clepsydra with i9nner points |  |
| 28 | 1 | Matra | 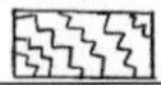 |
| | 2 | Matra with attached lines | 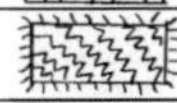 |
| | 3 | Matra with crenellated interior | 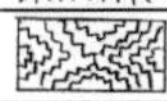 |
| | 4 | Matra with inner rectangle | 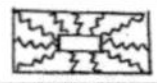 |
| | 5 | Matra with inner rhombuses | 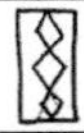 |
| 29 | 1 | Spot |  |
| | 2 | Spot with inner point |  |
| 30 | 1 | Crenallated cross and rhombus |  |
| 31 | 1 | Ñandú |  |
| 32 | 1 | Lion skin |  |
| | 2 | Others | |
| 33 | 1 | Rosette |  |
| | 2 | Aligned rosettes |  |
| 34 | 1 | Schematic lion |  |
| 35 | 1 | Guanaco |  |
| | 2 | Grouped guanacos |  |
| 36 | 1 | Human footprints |  |
| 37 | 1 | Human hands |  |
| | 2 | Negative human hands |  |

| | | | |
|---|---|---|---|
| 38 | 1 | Anthropomorphic figure |  |
| 39 | 1 | Contoured painted natural holes |  |
| 40 | 1 | Hollow star inscribed in dotted circle |  |
| | 2 | Filled star |  |
| 41 | 1 | Hook figure |  |
| 42 | 1 | Hollow triangle |  |
| | 2 | Filled triangle |  |
| | 3 | Dotted triangle | 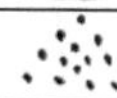 |
| | 4 | Triangle with a hook | 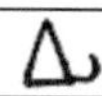 |
| 43 | 1 | T |  |
| | 2 | I |  |
| 44 | 1 | L |  |
| 45 | 1 | Arc and circle |  |
| 46 | 1 | Trident |  |
| 47 | 1 | Bola with handle | 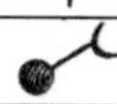 |
| 48 | 1 | Horse and rider |  |
| 49 | 1 | Horse |  |
| 50 | 1 | Y | 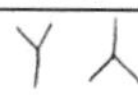 |
| 51 | 1 | Semicircle |  |
| 52 | 1 | Fusiform |  |
| | 2 | Fusiform with inner lines |  |
| 53 | 1 | Undulated line |  |
| | 2 | Parallel undulated lines | 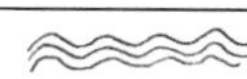 |
| 54 | 1 | 8 figure | 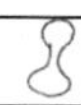 |
| | 2 | Axe |  |
| 55 | 1 | Trapeze | 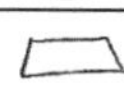 |
| | 2 | Linked trapezes | 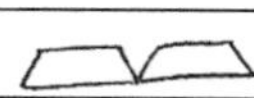 |
| 56 | 1 | Other zoomorphs | |
| 57 | 1 | Pyramid |  |
| 58 | 1 | E inverted |  |

# Appendix 3: Simulation

In order to discard the possibility of obtaining the values of Mutual Information of Table 2 by chance, we performed a simulation in which we randomly assigned the same number of 1s and 0s (the amount of "presences" of motifs in the sites) from the database in the $X_i$ variable. We generated 1000 random assignments. The obtained distribution of Mutual Information is shown in Fig. A3.1, where the threshold value *u* corresponds to a *p-value* (the probability of finding a case greater than the observed 0.093) of 1.56 % in the distribution of the Mutual Information obtained by random assignment. This means that, for the extreme case of less correlated pairs (on threshold value *u* 0.093), it is possible to obtain this correlation by random assignment with less than 1.56 (a low probability). Then, we performed another statistical test in which for each pair of motifs, the random assignment was made considering the same value as in our Table 2. For instance, comparing motif 50 with 15, we assigned 1 and 0 taking into account that motif 50 was present in 3 sites and motif 15 was present in 5 sites (see Table 2). We obtained a *p-value* less than 2 % for the three first cases corresponding to the threshold value *u* (the first three pairs of Table 2) and less than 0.2 % for the rest of the cases. Notice that these three cases, which are the ones with the greater possibility of being obtained by random, are also the ones with fewer information (given the small sample). Nevertheless we decided to include it on the Mutual Information Network, since in archaeology it is usual to deal with absence of information. In any case, with this exception, this test allows us to sustain that the probability of obtaining these values of Mutual Information correlation with the rest of the pair of motifs is low. These sets of correlated motifs above the threshold value *u* will be used to construct the MIN, in which nodes represent motifs and links represent the Mutual Information between them.

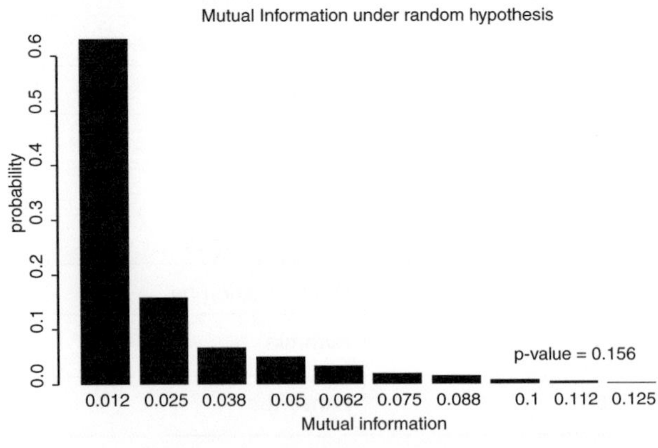


**Fig. A3.1** Histogram of Mutual Information values obtained by assigning the amount of 1s and 0s observed in the database randomly in a matrix of 49 columns and 43 rows. The *p*-value of 0.156 for *u* means that the probability of obtaining a value of Mutual Information greater than the threshold value is 1.56 %

# References


Albornoz, A. M. (1996). Sitios con arte rupestre en los alrededores del lago Nahuel Huapi (Río Negro). In J. Gómez Otero (Ed.), *Arqueología. Sólo Patagonia* (pp. 123–130). Puerto Madryn: Centro Nacional Patagónico.

Albornoz, A. M., & Cúneo, E. M. (2000). Análisis comparativo de sitios con pictografías en ambientes lacustres boscosos de Patagonia Septentrional: lagos Lácar y Nahuel Huapi (provincias de Neuquén y Río Negro). In M. M. Podestá & M. de Hoyos (Eds.), *Arte en las Rocas. Arte rupestre, menhires y piedras de colores en Argentina* (pp. 163–174). Buenos Aires: Sociedad Argentina de Antropología- Asociación Amigos del INAPL.

Arrigoni, G. (1997). Pintando entre lagos y bosques (las pinturas rupestres del Parque Nacional Los Alerces, Chubut). *Revista del Museo de Historia Natural de San Rafael, 16*(1–4), 241–259.

Arrigoni, G., & Fernández, P. (2004). Los restos óseos de alero del Sendero de Interpretación (PN Los Alerces, prov. del Chubut): integridad, resolución y aprovechamiento de los recursos faunísticos del bosque. In M. T. Civalero, P. M. Fernández, & A. G. Guráieb (Eds.), *Contra viento y marea. Arqueología de Patagonia* (pp. 403–416). Buenos Aires: Instituto Nacional de Antropología y Pensamiento Latinoamericano – Sociedad Argentina de Antropología.

Aschero, C. (1996). ¿Adónde van esos guanacos? In J. Gómez Otero (Ed.), *Arqueología. Solo Patagonia*. (pp. 153–152). Puerto Madryn: Centro Nacional Patagónico.

Aschero, C., Pérez de Micou, C., Onetto, M., Bellelli, C., Nacuzzi, L., & Fischer, A. (1983). *Arqueología del Chubut. El valle de Piedra Parada*. Rawson: Gobierno de la Provincia de Chubut.

Barrientos, G., & Perez, I. (2004). La expansión y dispersión de poblaciones del norte de Patagonia durante el Holoceno tardío: evidencia arqueológica y modelo explicativo. In T. Civalero, P. Fernández, & G. Guraieb (Eds.), *Contra Viento y Marea. Arqueología de Patagonia* (pp. 179–195). Buenos Aires: Instituto Nacional de Antropología y Pensamiento Latinoamericano & Sociedad Argentina de Antropología.

Barton, C., & Clark, G. A. (1997). Evolutionary theory in archaeological explanation. In C. Barton & G. A. Clark (Eds.), *Rediscovering Darwin: Evolutionary theory in archaeological explanation, archaeological papers of the American Anthropological Association* (pp. 3–18). Washington, DC: American Anthropological Association.

Barton, C. M., Clark, G. A., & Cohen, A. (1994). Art as information. Explaining Upper Palaeolithic art in Western Europe. *World Archaeology, 26*(2), 185–204.

Belardi, J. B. (2004). Más vueltas que una greca. In M. T. Civalero, P. M. Fernández, & A. G. Guráieb (Eds.), *Contra viento y marea. Arqueología de Patagonia* (pp. 591–603). Buenos Aires: Instituto Nacional de Antropología y Pensamiento Latinoamericano – Sociedad Argentina de Antropología.

Bellelli, C., Carballido Calatayud, M., Fernández, P. M., & Scheinsohn, V. (2003). El pasado entre las hojas. Nueva información arqueológica del noroeste de la provincia de Chubut, Argentina. *Werken, 4*, 25–42.

Binford, L. R. (1981). Behavioral archaeology and the Pompei Premise. *Journal of Anthropological Research, 37*, 195–208.

Borgerhoff Mulder, M., Nunn, C., & Towner, M. (2006). Cultural macroevolution and the transmission of traits. *Evolutionary Anthropology, 15*, 52–64.

Borrero, L. (1994–1995). Arqueología de la Patagonia. *Palimpsesto, 4*, 9–69.

Borrero, L., & Franco, N. (1997). Early Patagonian hunter-gatherers: Subsistence and technology. *Journal of Anthropological Research, 53*, 219–239.

Boschín, M. T. (2000). Sociedades cazadoras del área Pilcaniyeu, sudoeste de Río Negro: elementos para un análisis territorial. *Mundo Ameghiniano, 14*, 1–75.

Boschín, M. T. (2009). *Tierra de hechiceros. Arte indígena de la Patagonia septentrional argentina*. Salamanca, España: Ediciones Universidad de Salamanca. Servicio de publicaciones de la Universidad de Córdoba.

Boyd, R., & Richerson, P. (1985). *Culture and the evolutionary process*. Chicago: University of Chicago Press.

Carden, N. (2009). *Imágenes a través del tiempo. Arte rupestre y construcción social del paisaje en la Meseta Central de Santa Cruz*. Buenos Aires: Sociedad Argentina de Antropología.

Casamiquela, R. (1965). *Rectificaciones y ratificaciones hacia una interpretación definitiva del panorama etnológico de la Patagonia y área septentrional adyacente*. B. Blanca: Cuadernos del Sur, Instituto de Humanidades de la Universidad Nacional del Sur.

Castro, A. (2010). *Rutas Indígenas y Arqueología en la Provincia del Chubut*. Ph.D. manuscript, Buenos Aires: University of Buenos Aires.

Cavalli-Sforza, L. L., & Feldman, M. (1981). *Cultural transmission and evolution*. Princeton: Princeton University Press.

Chabris, C., & Simons, D. (2010). *The invisible gorilla and other ways our intuitions deceive us*. New York: Crown.

Claraz, G. (2008). *Viaje al río Chubut*. Buenos Aires: Ediciones Continente.

Cochrane, E. E., & Lipo, C. P. (2010). Phylogenetic analyses of Lapita decoration do not support branching evolution or regional population structure during colonization of Remote Oceania. *Philosophical Transaction of the Royal Society of London B Biology Science, 365*(1559), 3889–3902.

Conkey, M. (1978). Style and information in cultural evolution: Toward a predictive model for the Paleolithic. In C. Redman, et al. (Eds.), *Social archeology: Beyond subsistence and dating*. (pp. 61–85). New York: Academic Press.

Cover, T. M., & Thomas, J. A. (1991). *Elements of information theory*. New York: Wiley, Wiley Series in Telecommunications.

Crutchfield, J. P., Ellison, C. J., & Mahoney, J. R. (2009). Time's barbed arrow: Irreversibility, crypticity, and stored information. *Physical Review Letters, 103*, 094101.

Csardi, G., & Nepusz, T. (2006). The igraph software package for complex network research. *Inter Journal Complex Systems, 1695* (http://igraph.sf.net).

Millán de Palavecino, D. (1963). Área de expansión del tejido araucano. In *Primer Congreso del Área Araucana Argentina* (pp. 411–448) San Martín de los Andes: Junta de Estudios Araucanos.

Dowson, T. A. (1998). Revelations of religious reality: the individual in San rock art. *World Archaeology, 20*, 16–28.

Dunnell, R. C. (1978). Style and function: a fundamental dichotomy. *American Antiquity, 43*(1), 192–202.

Eerkens, J. (2000). Practice makes within 5% of perfect: The role of visual perception, motor skills, and human memory in artifact variation and standardization. *Current Anthropology, 41*(4), 663–668.

Escalada, F. (1949). *El complejo Tehuelche*. Buenos Aires: Coni.

Fiore, D. (2006). Poblamiento de imágenes: Arte rupestre y colonización de la Patagonia. Variabilidad y ritmos de cambio en tiempo y espacio. In D. Fiore & M. M. Podestá (Eds.), *Tramas en la piedra. Producción y usos del arte rupestre* (pp. 43–62). Buenos Aires: World Archaeological Congress, Asociación Amigos de Instituto Nacional de Antropología, Sociedad Argentina de Antropología.

Gamble, C. (1991). The social context for Palaeolithic art. *Proceedings of the Prehistoric Society, 57*(1), 3–15.

Gradin, C. (1999). Sobre las tendencias del arte rupestre de Patagonia argentina. In M. Tamagnini (Ed.), *Segundas Jornadas de Investigadores en Arqueología y Etnohistoria del Centro-Oeste del País* (pp. 85–99). Córdoba: Universidad Nacional de Río Cuarto.

Granovetter, M. (1973). The strength of weak ties. *American Journal of Sociology, 78*(6), 1360–1380.

Hajduk, A., Albornóz, A., & Lezcano M. (2004). El "mylodon" en el patio de atrás. Informe preliminar sobre los trabajos en el sitio El Trébol, ejido urbano de San Carlos de Bariloche. Provincia de Río Negro. In M. T. Civalero, P. M. Fernández, & A. G. Guráieb (Eds.), *Contra viento y marea. Arqueología de Patagonia* (pp. 715–732). Buenos Aires: Instituto Nacional de Antropología y Pensamiento Latinoamericano – Sociedad Argentina de Antropología.

Hajduk, A., Albornóz, A., & Lezcano M. (2009). Nuevas excavaciones en el sitio El Trébol (San Carlos de Bariloche, prov. de Río Negro). Más sobre los niveles con fauna extinta. In A. Austral & M. Tamagnini (Eds.), *Problemáticas en la arqueología contemporánea* (Vol. 3, pp. 955–966). Río Cuarto, Argentina: Universidad Nacional de Río Cuarto. Facultad de Ciencias Humanas. Facultad de Ciencias Exactas, Físico-Químicas y Naturales.

Henrich, J., Boyd, R., & Richerson, P. (2008). Five misunderstandings about cultural evolution. *Human Nature, 19*, 119–137.

Jordan P., & Shennan S. (2003). Cultural transmission, language, and basketry traditions amongst the California Indians. *Journal of Anthropological Archaeology, 22*, 42–74.

Kauffman, S. (1993). *The origins of order: Self-organization and selection in evolution*. Oxford: Oxford University Press.

Leclerc, R. (2008). Survival of the sparsest: Robust gene networks are parsimonious. *Molecular Systems Biology, 4*, 213.

Lewis-Williams, J., & Dowson, T. (1988). The sign of all Times: Entoptic Phenomena in Upper palaeolithic Art. *Current Anthropology, 29*(2), 201–245.

Loecher, M. (2012). RgoogleMaps: Overlays on Google map tiles in R. R package version 1.2.0.2. http://CRAN.R-project.org/package=RgoogleMaps. Berlin School of Economics and Law.

Mackay, D. (2003). *Information theory, inference and learning algorithms*. Cambridge: Cambridge University Press.

McDonald, J., & Veth, P. (2011). Information exchange amongst hunter-gatherers of the Western Desert of Australia. In R. Whallon, W. Lovis, & K. Hitchcock (Eds.), *Information and its role in hunter-gatherer bands* (pp. 221–223). Los Angeles: Cotsen Institute of Archaeology Press.

Menghin, O. (1957). Estilos del arte rupestre de Patagonia. *Acta Praehistorica, I*, 57–87.

Mesoudi, A. (2008). The experimental study of cultural transmission and its potential for explaining archaeological data. In M. J. O'Brien (Ed.), *Cultural transmission and archaeology: Issues and case studies* (pp. 91–101). Washington, DC: Society of American Archaeology Press.

Mitchell, M. (2009). *Complexity: A guided tour*. Oxford: Oxford University Press.

Mithen, S. (1988). Looking and learning: Upper Palaeolithic art and information gathering. *World Archaeology, 19*, 297–327.

Morello, J. (1984). *Perfil Ecológico de Sudamérica. Características estructurales de Sudamérica y su relación con espacios semejantes del planeta*. Barcelona: ICI – Ediciones Cultura Hispánica.

Moreno, F. (1997). *Reminiscencias del Perito Moreno*. Buenos Aires: Elefante Blanco.

Neiman, F. (1995). Stylistic variation in evolutionary perspective: Inferences from decorative diversity and interassemblage distance in Illinois woodland ceramic assemblages. *American Antiquity, 60*(1), 7–36.

O'Brien, M. J., & Lyman L. R. (2003). Introduction. In M. J. O'Brien, & L. R. Lyman (Eds.), *Style, function, transmission: Evolutionary archaeological perspectives* (pp. 1–32). Salt Lake City: University of Utah Press.

Onetto, M. (1986–1987). Nuevos resultados de las investigaciones de Campo Nassif 1. Valle de Piedra Parada. Provincia del Chubut. *Relaciones de la Sociedad Argentina de Antropología, 17*(1), 95–123.

Pedersen, A. (1978). Las pinturas rupestres del Parque Nacional Nahuel Huapi. *Anales de Parques Nacionales, XVI*, 7–43.

Pérez de Micou, C. (1979–1882). Sitio Piedra Parada 1 (PP1), departamento Languiñeo, provincia de Chubut (República Argentina). *Cuadernos del Instituto Nacional de Antropología y Pensamiento Latinoamericano, 9*, 97–111.

Pérez de Micou, C., Trivi de Mandri, M., & Burri, L. (2009). *Imágenes desde un alero. Investigaciones multidisciplinarias en Río Mayo, Chubut. Patagonia argentina*. Buenos Aires: Fundación de Historia Natural Félix de Azara.

Podestá, M., Bellelli, C, Labarca, R. Albornoz, A. M., Vasini, A., & Tropea, Y. E. (2008). Arte rupestre en pasos cordilleranos del bosque andino patagónico (El Manso, Región de los Lagos y Provincia de Río Negro, Chile-Argentina). *Magallania, 36*, 143–153.

Podestá, M., Albornoz, A. M., Vasini, A., & Tropea, E. (2009). El sitio Peumayén 2 en el contexto del arte rupestre del bosque andino patagónico. *Comechingonia Virtual, 3*(2), 117–153. http://www.comechingonia.com. Accessed 6 June 2014.

R Development Core Team. (2008). R: A language and environment for statistical computing. R Foundation for Statistical Computing, Vienna. http://www.R-project.org.
Richerson, P., & Boyd, R. (2005). *Not by genes alone: How culture transformed human evolution*. Chicago: University of Chicago Press.
Ross, C., & Richerson, P. (2014). New frontiers in the study of cultural and genetic evolution. *Current Opinion in Genetics and Development, 29*, 103–109.
Scheinsohn, V. (2003). Hunter gatherer archaeology in South America. *Annual Review of Anthropology, 32*, 339–361.
Scheinsohn, V. (2011). Rock art information among hunter-gatherers in Northwest Patagonia: an assessment of broad-scale and territorial models. In R. Whallon, W. Lovis, & K. Hitchcock (Eds.), *Information and its role in hunter-gatherer bands* (pp. 235–247). Los Angeles: Cotsen Institute of Archaeology Press.
Scheinsohn V., & Matteucci, S. (2013). A regional model of archaeological Distributions for Northwestern Andenan Patagonia (Argentina). In M. Figuerero Torres & A. Izeta (Eds.), *El uso de Sistemas de Información Geográfica (SIG) en la arqueología sudamericana*. (pp. 61–72). Oxford: BAR International Series 2497.
Scheinsohn, V., Szumik, C., Leonardt, S., & Rizzo, F. (2009). Distribución espacial del arte rupestre en el bosque y la estepa del norte de Patagonia. Nuevos resultados. In M. Salemme, F. Santiago, M. Álvarez, E. Piana, M. Vázquez, & F. Mansur (Eds.), *Arqueología de la Patagonia. Una mirada desde el último confín* (pp. 541–558). Ushuaia: Editorial Utopías.
Scheinsohn, V., Szumik, C., Leonardt, S., & Rizzo, F. (2015). *The "Hidden" Code: Coding and classifying in rock art: A Northwestern Patagonia case study*. Journal of Archaeological Method and Theory, pp 1–20. First online: 03 May 2015. DOI 10.1007/s10816-015-9249-8.
Schiffer, M. B. (1972). Archaeological context and systemic context. *American Antiquity, 37*(2), 156–165.
Shannon, C. (1948). A mathematical theory of communication. *Bell System Technical Journal, 27*; 379–423.
Shennan S., & Bentley, R. (2008). Style, interaction, and demography among the earliest farmers of Central Europe. In M. O'Brien (Ed.), *Cultural transmission and archaeology: Issues and case studies* (pp. 164–177). Washington, DC: Society of American Archaeology Press.
Shennan, S., & Wilkinson, R. (2001). Ceramic style change and neutral evolution: A case study from Neolithic Europe. *American Antiquity, 66*(4), 577–593.
Silveira, M. (1988–1989). Un sitio con arte rupestre: el Alero Lariviere (provincia de Neuquén). *Relaciones de la Sociedad Argentina de Antropología, 17*(2), 75–86.
Silveira, M. (1996). Alero Los Cipreses (Provincia del Neuquén, República Argentina). In J. Gómez Otero (Ed.), *Arqueología Solo Patagonia* (pp. 107–118). Puerto Madryn, Argentina: Centro Nacional Patagónico – Consejo Nacional de Investigaciones Científicas y Técnicas.
Silveira, M. (1999). El Alero Lariviere: un sitio en el bosque septentrional andino (Departamento Los Lagos, Provincia de Neuquén, Argentina). In J. B. Belardi, P. Fernández, R. Goñi, G. Guráieb, & M. De Nigris (Eds.), *Soplando en el Viento. Actas de las III Jornadas de Arqueología de la Patagonia* (pp. 83–92). Neuquén, Argentina: Instituto Nacional de Antropología y Pensamiento Latinoamericano. Universidad Nacional del Comahue, Facultad de Humanidades.
Silveira, M. J., & Fernández, M. B. (1991). El arte rupestre de la cuenca del Lago Traful (Provincia del Neuquén). In M. Podestá, M. I. Hernández Llosas, & S. F. Renard de Coquet (Eds.), *El arte rupestre en la arqueología contemporánea* (pp. 101–112). Buenos Aires: SF.
Simonetti, F., Teppa, E., Chernomoretz, A., Nielsen, M., & Busljе, C. (2013). MISTIC: Mutual information server to infer coevolution. *Nucleic Acids Research, 41*, Web Server issue.
Sperber, D. (1996). *Explaining culture: A naturalistic approach*. New York: Blackwell.
Stark, M. T., Bowser, B. J., & Horne, L. (Eds.). (2008). *Cultural transmission and material culture*. Tucson: The University of Tucson Press.
Tehrani J., & Collard, M. (2002). Investigating cultural evolution through biological phylogenetic analyses of Turkmen textiles. *Journal of Anthropological Archaeology, 21*, 443–463.

Torkamaniet, A., Dean, B., Schork, N. J., & Thomas, E. A. (2010). Coexpression network analysis of neural tissue reveals perturbations in developmental processes in schizophrenia. *Genome Research, 20*, 403–412.

Vignati, M. A. (1944). Antigüedades en la Regi6n de los Lagos Nahuel Huapi y Traful 1- 7. *Notas del Museo de La Plata, IX*(23), 24–29.

Watts, D. J., & Strogatz, S. H. (1998). Collective dynamics of small-world networks. *Nature, 393*, 440–442.

Whallon, R. (2006). Social networks and information. Non-"utilitarian" mobility among hunter-gatherers. *Journal of Anthropological Archaeology, 25*, 259–270.

Whallon, R. (2011). An introduction to Information and its role in Hunter-Gatherer band-level Societies. In R. Whallon, W. Lovis, & R. Hitchcock (Eds.), *Information and its role in Hunter-Gatherer Bands* (pp. 1–18). Los Angeles: Cotsen Institute of Archaeology Press, University of California.

White, A. (2013). An abstract model showing that the spatial structure of social Networks affects the outcomes of cultural transmission processes. *Journal of Artificial Societies and Social Simulation, 16*(3), 9. http://jasss.soc.surrey.ac.uk/16/3/9.html. Accessed 03/02/2015.

Whitley, D. (2005). *Introduction to rock art research*. Arizona: Left Coast Press.

Whitley, D. (2011). Rock art, religion, and ritual. In T. Insoll (Ed.), *Oxford handbook of the archaeology of ritual and religion* (pp. 307–326). New York: Oxford University Press.

Wobst, H. M. (1977). Stylistic behavior and information exchange. In C. Cleland (Ed.), *Papers for the director: Research essays in honor of James B. Griffin. Anthropological Papers 61* (pp. 317–342). Ann Arbor: University of Michigan Museum of Anthropology.

# A Cladistics Analysis Exploring Regional Patterning of the Anthropomorphic Figurines from the Gravettian

**Allison Tripp**

**Abstract** Numerous studies have interpreted the anthropomorphic "Venus" statuettes of the Gravettian. However, few of these studies have scrutinized the figurines at an individual level or used quantitative analyses in order to understand similarities within sites or between regions. This study tests two hypotheses. The first one, by Leroi-Gourhan, suggests that the Gravettian statuettes share core similarities regardless of where they were created. If correct, statuettes should not be grouped according to the region that they were made. The second hypothesis, by Gvozdover, suggests a Kostenki-Avdeevo unity. Her hypothesis suggests blending among cultures in the Russian Plains and that there are "types" of statuettes that are not restricted to a particular site. Here cladistics methods are used in order to understand whether ethnogenesis (blending) or cladogenesis (branching) has occurred in the production of "Venus" making. Results confirm and extend Gvozdover's hypothesis suggesting cultural and ideological connections for "Venus" making in the Russian Plains and also support the uniqueness of a few European statuettes.

**Keywords** paleolithic art • Venus • anthropomorphic • figurines • Gravettian

## 1 Introduction

For more than 100 years, archaeologists and historians have tried to understand a unique group of Paleolithic objects (White 2006). These artifacts include bas reliefs, miniature masks, pendants (Fig. 1),[1] figurines, miniature heads, and ambiguous "sexual" objects.[2] More than 50 such figurines, frequently termed "Venuses," were

[1] "Pendant" is often used to cover statuettes that are less than 100 mm in length and less than 30 mm in width (Hahn 1990).

[2] See Fig. 1 for an example. This group contains a variety of objects that have been labeled "Venuses." This includes depictions of sexual organs, breasts, or objects that arguably are not even

A. Tripp
Department of Anthropology, Chaffey College, Rancho Cucamonga, CA, USA
e-mail: allisonjtripp@gmail.com

 
L. Mendoza Straffon (ed.), *Cultural Phylogenetics: Concepts and Applications in Archaeology*, Interdisciplinary Evolution Research,
DOI 10.1007/978-3-319-25928-4_8

**Fig. 1** Cast of "Venus" XIII discovered in 1935 and composed of ivory, Pavlovian (Photo by A. Tripp at Monrepos, cast no. unavailable, Verpoorte 2001: 46. Leroi-Gourhan's (1968) "lozenge composition demonstrating commonalities in the representation of the female form)

discovered at sites from France to Siberia and as far south as Italy and are dated to the Gravettian (Fig. 2 – the map on Fig. 2 was made by Sean Dolan). Chronometrically, this period begins and ends at different times across Europe. Its earliest appearance is in Eastern Europe around 30,000 BP and its latest occurrence is in Italy around 16,000 BP (Svoboda 2000; Pettitt 2000). While these dates may be an accurate reflection of former occupations, it may also reflect either contaminated samples or problems with the calibration curve (Pettitt 2000). In this study, statuettes that were radiometrically or stylistically dated to this techno-complex were included in the analysis. Although the Gravettian is not the earliest occurrence for anthropomorphic statuettes, only one has been found from the preceding period, the Aurignacian.

---

anthropomorphic.

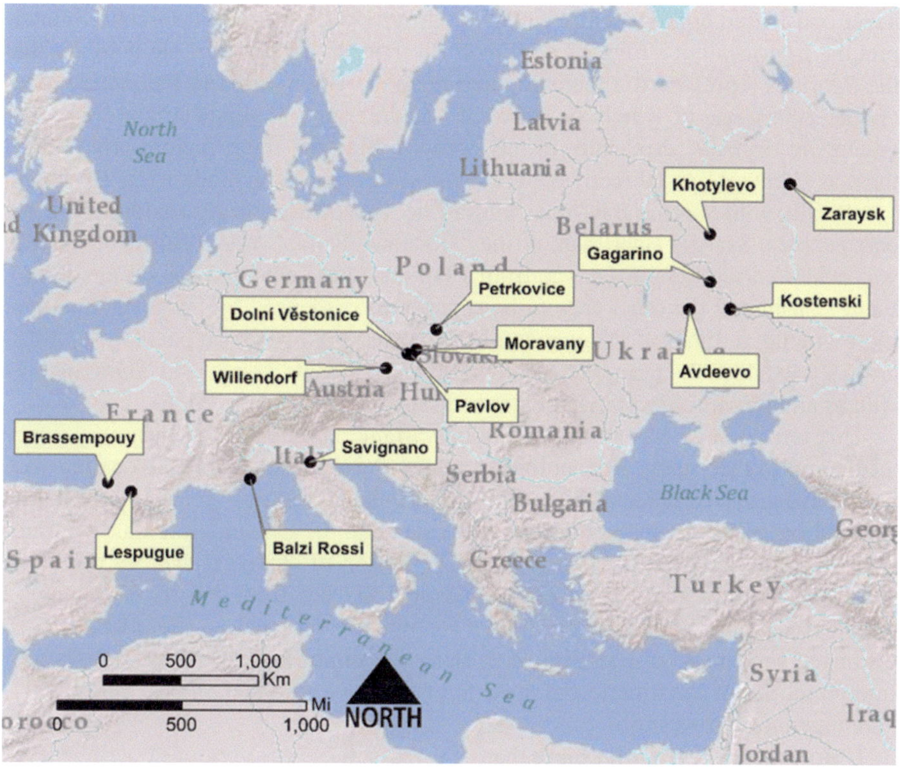


**Fig. 2** A map of Gravettian sites yielding female figurines. Map created by Sean Dolan

Since the majority of the statuettes is female and is nude, this has led many researchers to suggest that the figurines were created for a uniform same purpose (Soffer and Praslov 1993). Early interpretations include to guard property (Von Koeningswald 1972), to promote alliance networks (Gamble 1982), for use as teaching or obstetrical aides (McCoid and McDermott 1996), for fertility magic (Reinach 1903, Count Bégouen 1925), to represent Paleo-erotica (Guthrie 2005), or to represent a shared belief in a mother goddess (Hawkes and Wolley 1963; Levy 1948; Markale 1999; Hawkes and Wolley 1963). However, few studies (see Gvozdover 1995 for an exception) analyze the statuettes at an individual level and instead simply promote a hypothesis based on assumed stylistic similarities.

At present, there has not been a single study demonstrating that the aforementioned artifacts are a homogenous group based on any particular combination of shared stylistic features. The figurines are only assumed to represent a cohesive unit because many of the statuettes are nude and female and have exaggerated sexual features. This however masks the diversity that is apparent when examining individual figurines. Not all figurines are female, many appear to be wearing clothing, and a diversity of body shapes is also apparent (apple, pear, reversed triangle). In fact, from the discovery and initial interpretation of Gravettian female statuettes, they were never considered homogeneous. French prehistorian Edouard Piette sought to interpret the figurines but struggled with their diversity in appearance. He believed that the figurines represented a realistic interpretation of two different Paleolithic races (White 2006), one of which, he believed, was inferior and characterized by having greater fat deposits, especially in the stomach, hip, and thigh regions. He labeled these as "Venuses" and directly connected these figurines with modern day Bushman, specifically with Saartjie Baartman. He saw her as a contemporary analogue of these past peoples. Saartjie Baartman, or the "Hottentot Venus," was a Bushwoman who was exhibited throughout Europe from 1810 until her death in 1816 (White 2006). Piette considered the other race to be superior as evidenced by figurines that were more gracile and lacked exaggerated fat deposits. Following, archaeologists indiscriminately added the term "Venus" to figurines, pendants, beads, and engravings, as well as ambiguous objects that are not anthropomorphic (White 2006).

This terminology has caused much unwarranted confusion because it assumes affiliation. In fact, many archaeologists simply presuppose that we are dealing with a cohesive group of objects, since they are all called "Venuses" (Nelson 1990). The application of the term also reinforced the idea that the Gravettian female figurines share a single function (Nelson 1990). The notion of shared usage did not come from microscopic analysis or contextual analysis but from the term "Venus." I would argue that if these objects did not share that name, they would not have been evaluated as a single unit or argued to share the same function. Over time various interpretations of the function of these figurines have emerged all based on Piette's "homogenous" grouping.

Unfortunately, even after more than 100 years, these assumptions continue to be made. Conard (2009: 251), for example, argued "although there is a long history of debate over the meaning of Paleolithic Venuses, their clearly depicted sexual attributes suggest that they are a direct or indirect expression of fertility." By focusing only on exaggerated sexual features, we are ignoring countless other variables that are expressed on the figurines. One could argue that this is a result of a male bias on the interpretation of these artistic objects. We are also projecting our modern ideals of nudity and exaggerated sexual features as being connected to fertility. Again, interpretations like this are problematic because they assume affiliation of these anthropomorphic images and are also untestable.

A few scholars have argued that these objects represent a diversity of forms and therefore variable functions, albeit with limited quantitative data (e.g., see Nelson 1990; Soffer 1987; Rice 1981; Ucko 1962). Some researchers have documented diversity in morphology. For example, Gvozdover (1995) found differences in the location of ornamentation on the statuettes depending on their region of origin. For example, figurines from West Europe display decorations on the hips, thighs, and occasionally the breast, while those from Central and Eastern Europe emphasize the stomach and breasts. Gvozdover (1989) also demonstrated that certain patterns that were found on the figurines from the Russian Plains were also produced on objects that are not anthropomorphic, which she referred to as "shovels." Duhard (1993) argued that the figurines represented faithful depictions of Paleolithic females because they represented modern body types including steatopygia (having excessive gluteal fat), steatocoxia (fat around the hips), steatotrochanteria (femoral fat), and steatomeria (crural fat). As a gynecologist, he determined which of the figurines were pregnant and which were not. He concluded that 68 % of the Gravettian statuettes appear pregnant in comparison to 36 % in the Magdalenian (Duhard 1993). He argued that more could represent pregnancy because they have their arms directed to their abdomen (Duhard 1993). But, it is important to remember that while a figurine may appear pregnant, that might not be related to the function or the meaning of the object to Paleolithic people. For this reason, it is important to focus on multiple variables instead of just one, when analyzing the figurines. I have argued that there is evidence for both homogeneity and heterogeneity in the waist-to-hip ratios (WHR) of the Gravettian female figurines depending on their region of origin and that the resulting range in the WHR represents women in all phases of life (Tripp and Schmidt 2013). At present, a comprehensive quantitative multiregional review analyzing the individual details of the Gravettian "Venuses" is lacking.

In the present paper, I will argue that in order to understand regional connections, we first need to look at individual differences. To investigate this, I will use cladistics to test two contrasting hypotheses, one by Leroi-Gourhan (1982) and the other by Gvozdover (1989). Upon surveying the corpus of Gravettian anthropomorphic figurines, Leroi-Gourhan (1968: 96) wrote:

> No matter where found … they are practically interchangeable, apart from their proportions. The most complete figures have the same treatment of the head, the same small arms folded over the breasts or pointing towards the belly, the same low breasts drooping like sacks to far below the waist, and the same legs ending in miniscule or non-existent feet.

This is referring to Leroi-Gourhan's description of the "lozenge composition," which showed statuettes from all regions except Siberia (Fig. 3). In general, the lozenge shape can be reflective of the contours of a woman's body. There is a fanning out from the shoulders to the hips and a narrowing toward the feet. This is in contrast to a male body, which would generally be broad at the shoulders and then be narrow toward the feet, forming an upside-down triangle. What is interesting about the drawing by Leroi-Gourhan is that the widest part of the lozenge is the waist and not the hips; this either represents pregnancy or severe obesity. After puberty, most women, unlike men, store the majority of their body fat in their hips and thighs not in their abdominal region (Buss 2004).

Leroi-Gourhan's (1968) statement reflects the idea that Gravettian anthropomorphic figurines are more similar than different and that no matter where they were created, they share core features. This assumption is shared by several scholars including Gamble (1982), McCoid and McDermott (1996), McDermott (1996), Taylor (1996), Guthrie (2005), and Conard (2009). In contrast, Gvozdover (1995) has argued that statuettes from the Russian Plains including Kostenki, Avdeevo, Gagarino, and Khotylevo are stylistically similar and represent a united cultural group, presumably to the exclusion of statuettes from other regions.

Both of these hypotheses can be tested through the use of a cladistics analysis and each would lead to the creation of different diagrams. Leroi-Gourhan's hypothesis, henceforth referred to as hypothesis 1, suggests that regardless of their

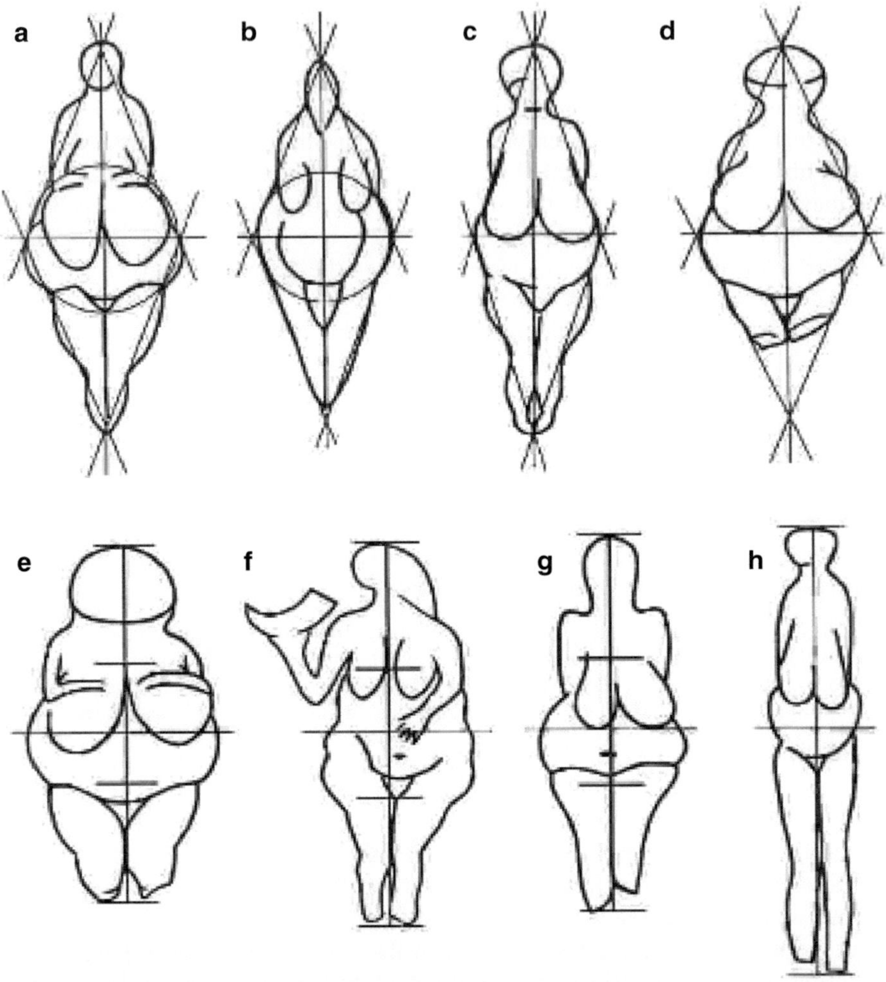


**Fig. 3** Leroi-Gourhan's (1968) "lozenge composition demonstrating commonalities in the representation of the female form

location, they share major features. This would suggest that statuettes from each region would be spread across the tree indiscriminately, i.e., there would be no clear geographic clustering in the dataset. This would support Leroi-Gourhan's hypothesis as he does not suggest that statuettes from one region, or site, share more similarities with each other than those from another group. Hypothesis 2, by Gvozdover (1989), argues for similarities in artistic industries among the sites in the Russian Plains including Kostenki I, Avdeevo, New Avdeevo, Gagarino, and Khotylevo II. More specifically, she argued for the presence of Kostenki, Avdeevo, and Gagarino-type statuettes but suggested that they were not limited to a single site. This hypothesis suggests contact and blending of Russian artistic cultures. If this idea is correct, then individual statuettes might not group according to their site but should cluster into a monophyletic Russian group to the exclusion of statuettes from other regions.

By exploring regional patterning, this analysis will also provide insight into cultural evolution. Specifically, it will clarify whether vertical (branching) or horizontal (blending) transmission had a greater effect on "Venus" making in the Gravettian. Cultural phylogenesis is analogous to biological phylogenesis and refers to the evolution of culture through the progressive splitting of cultural assemblages (Tehrani and Collard 2002). In cultural phylogenesis, information is passed from parent to offspring within a single culture, referred to as vertical transmission (O'Brien and Lyman 2003). Differences between groups would suggest independent development of artistic traditions at a particular site or within a region. Ethnogenesis is when cultural evolution occurs through the exchange of ideas, beliefs, practices, etc., between contemporaneous populations and is analogous to biological gene flow (Tehrani and Collard 2002). This occurs when individuals from different populations are intermarrying, copying each other's ideas, as well as exchanging ideas (Collard et al. 2006) – i.e., through the process of horizontal transmission of culture. Close similarity between statuettes from different sites/regions could provide evidence for the horizontal exchange and blending of artistic traditions, while distinct differences among groups or clusters of groups would favor a branching (vertical) explanation for cultural change.

## 2 Methodological Background

### 2.1 *Cladistics*

A phylogeny is "the evolutionary history of a group of organisms" (Campbell et al. 2003:295). Cladistics is based on the Darwinian concept of evolution, specifically descent with modification (Jordan and Shennan 2003). This idea suggests that new species arise from existing ones as they change and adapt over time. A phylogenetic tree (cladogram) is created by grouping together species with the most shared similarities. By studying the tree, one can learn about the evolutionary relationships between the taxa involved. Cladistics can be used to study classification and diversity.

This study will employ cladistics analyses in order to examine diversity and regional patterning among the statuettes. The first step is selecting an out-group, usually a close relative to the group, and generally the artifact or species that deviated earliest (the oldest) (Holden and Shennan 2005). An out-group is important because it is assumed to represent the ancestral character states. By comparing the remaining taxon to the out-group, it is possible to analyze which character states changed and when these changes occurred. This analysis, for example, uses the "Venus" of Hohle Fels as an out-group because it is the oldest anthropomorphic figurine and is from a related archaeological culture, the Aurignacian. Afterward, a data sheet of the character states is created for all of the taxa (specimens). Two strengths of this method are that it allows for a multivariable analysis of the figurines and weighs each variable equally. As previously mentioned, too many hypotheses have focused only on sexual features and have ignored other diagnostic features. By analyzing numerous features, I will be able to make more meaningful statements about the overall similarity and differences that we see among the figurines and how this relates to group or individual identity.

Once the data matrix is created, the data can be run. The principle of parsimony is used to construct the tree. Buchanan and Collard describe the process as the following: "when a character occurs in two states among the study group, but only one of the states is found in the outgroup, the principle of parsimony is invoked and the state found only in the study group is deemed to be evolutionarily novel with respect to the outgroup state" (Buchanan and Collard 2007, 368). This means that the simplest explanation is preferred in order to construct the tree, for example, taking one evolutionary step, instead of four. This leads to fewer assumptions and clearer resolution within the tree (O'Brien and Lyman 2003).

When analyzing the data, it is important to recognize homologies and analogies. Homologies are the result of similarities due to shared ancestry. For example, dogs and mice share fur and milk ducts because their common ancestor did. Homologies are thus useful because they provide information about a shared evolutionary heritage. When a character is found among two groups that are not related, this is known as an analogy and can result from two processes (O'Brien and Lyman 2003). One of which is parallel (convergent) evolution of a trait. In this scenario, two unrelated species adapted to a similar environment in a similar way and both develop the character state(s) (i.e., flying in both birds and butterflies). In the current study, an example of this would be equivalent to figurines from unrelated sites grouping together because they shared decorations on their legs and arms. Decorations on the front and back of the torso were relatively common and were thus distinguished by different character states such as incision, puncture, and rope. However, decorations on the arms and legs were uncommon and were coded for presence or absence. For this reason, if particular statuettes did not have enough unique features to differentiate themselves as a group, they could be most closely related to another statuette that did have decorations on the arms and legs. In this case, we would say that the statuettes were not pairing together due to communication and the copying of traits among the groups but were the result of independent innovation. The other process leading to analogy is a character state reversal (O'Brien and Lyman 2003). This last process is not seen in the current study. The presence of analogies is problematic because they can lead to multiple versions of trees. Parsimony is applied and the shortest tree is used.

The incorporation of the data allows for the construction of a branching diagram, demonstrating the possible relations between the taxa. The tree traces how the specimens diverged from the out-group and which individuals are related to each other based on the most shared similarities.

## 2.2 Cladistics and Archaeological Applications

Archaeologists have long noted that typological similarity among objects is connected to cultural relatedness, but the application of cladistics to understanding cultural evolution is relatively new. Like biological traits, cultural ideals and practices evolve too. Our knowledge of how populations change biologically over time, however, is much better understood than how they change culturally. Shennan (2008, p. 76) describes cultural evolution as "the changing distributions of cultural attributes in populations, likewise affected by processes of natural selection but also by others that have no analog in genetic evolution." While it may be argued that processes of cultural evolution differ from biological evolution, there are many important similarities. In both cases, genes, or artifacts in this case, are modified over time. In both cases, individuals are the focus of selection, but it is the population that evolves. In a biological sense, individuals that are the fittest pass on their genes and through natural selection, the population evolves. In a cultural sense, the individual passes on information about how an artifact is made or how it should look. That knowledge then becomes a part of the cultural gene pool, which can be modified by the community. Additionally, Shennan (2008) argued that cultural traits may provide an advantage to those that acquire them and that this may allow the individual a greater chance at survival and reproduction. Thus, cultural traits, like biological traits, can also evolve through natural selection. An important difference is that cultural evolution can progress much more quickly than biological evolution (O'Brien et al. 2001). Others however have questioned whether or not using a biological method to analyze cultural evolution is appropriate (see Bamforth 2002; Fracchia and Lewontin 1999, 2005). A detailed discussion on this topic however is beyond the scope of this paper.[3]

Although cladistics methodologies were originally developed to demonstrate associations between biological species, more recently they have been applied to other fields including paleontology, botany, zoology, linguistics, and archaeology. Tehrani and Collard, for example, used cladistics to evaluate whether vertical versus horizontal transmission had a stronger role in the creation of Turkman woven artifacts (Tehrani and Collard 2002). Buchanan and Collard (2007) examined Paleo-Indian projectile points and used phylogenetics to test various hypotheses for dispersal patterns for the colonization of North America. Rexova and colleagues (2003) investigated whether borrowing or branching occurred more frequently among languages from the Indo-European language family. Others have utilized this method to investigate various forms of material culture including basketry, lithics, and pottery to test predictions about cultural variation (Collard and Shennan 2000; Jordan and Shennan 2003; Jordan 2009).

The majority of the previously mentioned studies have been done on artifacts with functional uses. The figurines utilized in the current study may have had many uses and may be in various stages of completion. The statuettes exhibit variation in

[3] For further information on the appropriateness of using cladistics for testing hypotheses on cultural evolution, refer to O'Brien and colleagues (2001), Eerkens and colleagues (2006), and Shennan (2008).

every part of their bodies. This may be related to the desire of the maker or the culture, the function of the object, the skill of the maker, or whether or not it was finished. Additionally, they may also display convergent similarities, because there are only so many ways to create a realistic anthropomorphic object. Therefore, the material culture analyzed here may be much more difficult to interpret.

## 3 Method

The sample size for this analysis was composed of 30 discrete traits of equal weight found on 27 statuettes. Only statuettes that were complete and anthropomorphic (i.e., clearly human-like) in nature were included in the study. Eleven casts held at Schloss Monrepos Museum in Germany were analyzed directly. One artifact was studied at the Kunstkamera, located in St. Petersburg, Russia. The others were studied using published drawings and photographs (Delporte 1993; Ambramova 1995; Gvozdover 1995; Mussi et al. 2000). Each of the previously mentioned statuettes was examined from the dorsal, ventral, and lateral perspectives.

The 39 traits were found throughout the body. This included the location and types of decoration as well as features related to anatomical details on the figurines (e.g., profile head shape, torso proportion). Each figurine was analyzed using the character states found in Appendix 1. The data were then put into a character state data matrix. TNT (Tree Analysis Using New Technology) was then used to run the data (Goloboff et al. 2000), and Dendroscope was used to create the final image (Hudson and Scornavacca 2012). TNT was used to test both hypotheses because it creates unbiased groups based on shared features. Appendix 2 presents the character states that are less straightforward. Table 1 lists the statuettes utilized, site, region, collection, and source (i.e., photos or original specimens).

Only figurines that were complete, or nearly complete, were utilized for the analysis, as a trial run including incomplete specimens did not yield fruitful results (i.e., statuettes were grouped together because they share missing arms, legs, and heads and not necessarily because they share stylistic similarities). The sample is therefore biased toward figurines from the Russian Plains. This is due to the fact that many more figurines were created at those sites but also because the majority is complete. This is in contrast to sites like Dolní Věstonice, where the many anthropomorphic ceramic statuettes are highly fragmented. In fact, Soffer and colleagues (1993) found that of the 720+ figurines, 99.9 % are broken, and the damage was ancient, likely due in part to firing temperatures (Soffer and Praslov 1993).

The data were analyzed to find the most parsimonious tree. A basic analysis was conducted using TNT with the Venus of Hohle Fels as the out-group. Branch swapping was performed after multiple addition sequences were used. This analysis is comparable to a "heuristic search" with random addition sequences in PAUP* or hold/10; mult*10; in NONA (Goloboff et al. 2008). The result was two trees, so a consensus tree was created. Another statistical measure was employed to analyze how well the data fit the tree. The retention index (Farris 1989) divides a fraction of

**Table 1** Details of figurines employed in analyses

| Specimen # or name | Site | Region | Source |
|---|---|---|---|
| Abrachial | Grimaldi/Balzi Rossi | S. Europe | Photo analysis |
| Avdeevo 48–3 | Old Avdeevo | E. Europe | Photo analysis |
| Avdeevo 48–5 | Old Avdeevo | E. Europe | Monrepos |
| Avdeevo 76–6 | New Avdeevo | E. Europe | Photo analysis |
| Avdeevo 77–1 | New Avdeevo | E. Europe | Photo analysis |
| Avdeevo 78–9 | New Avdeevo | E. Europe | Photo analysis |
| Bicephalous | Grimaldi/Balzi Rossi | S. Europe | Photo analysis |
| Dolní Věstonice Venus | Dolní Věstonice I | E. Europe | Monrepos |
| Gagarino 2 | Gagarino | E. Europe | Monrepos |
| Gagarino 3 | Gagarino | E. Europe | Monrepos |
| Hohle Fels | Hohle Fels | E. Europe | Photo analysis |
| Khotylevo II-2 | Khotylevo II | E. Europe | Photo analysis |
| Khotylevo II-3 | Khotylevo II | E. Europe | Photo analysis |
| Kostenki male | Kostenki I | E. Europe | Photo analysis |
| Kostenki 83-I | Kostenki I | E. Europe | Photo analysis |
| Kostenki 3 | Kostenki I | E. Europe | Kunstkamera/Monrepos |
| Kostenki 4 | Kostenki I | E. Europe | Monrepos |
| La fillette | Brassempouy | W. Europe | Monrepos |
| Lespugue | Lespugue 19 | W. Europe | Monrepos |
| Lozenge | Grimaldi/Balzi Rossi | S. Europe | Photo analysis |
| Moravany | Moravany-Podkovica | C. Europe | Monrepos |
| Pavlov Venus | Pavlov I | C. Europe | Monrepos |
| Punchinello | Grimaldi/Balzi Rossi | S. Europe | Photo analysis |
| Savignano | Savignano | S. Europe | Photo analysis |
| Willendorf (Venus) | Willendorf II | C. Europe | Monrepos |
| Zaraysk Venus | Zaraysk | E. Europe | Photo analysis |

possible homoplasy by the maximum possible value of homoplasy. This measure is often used instead of the consistency index, as it is less influenced by the number of taxa, characters, and character states in the data (Archie 1989). Values of the RI range from 0, indicating complete homoplasy, to 1, indicating no homoplasy in the data. According to Nunn et al. (2010), values above a 0.6 indicated high levels of branching and hence low levels of ethnogenesis.

## 4 Results

The final results can be visualized in the cladogram (Fig. 4). The retention index associated with the most parsimonious cladogram was 0.443. As the cladogram shows, several figurines do not share direct associations with other statuettes

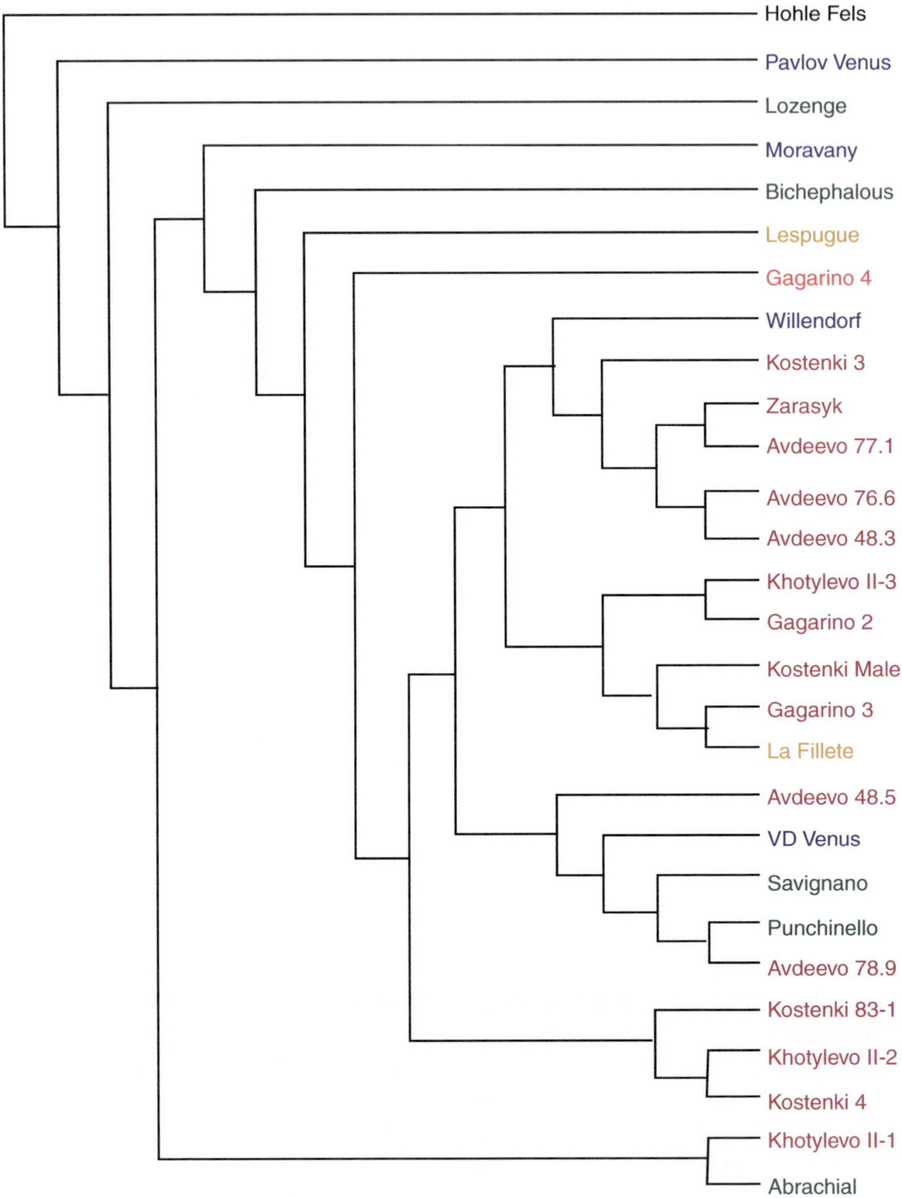


**Fig. 4** The final cladogram of 27 anthropomorphic statuettes from the Gravettian. The retention index was 0.443

including the Venus of Pavlov, the Lozenge, Moravany, the Bicephalous, and Lespugue. For the most part, the Russian Plains statuettes group together, while the Central, Southern, and Western European statuettes groups spread throughout the tree.

## 5 Discussion

The results do not support hypothesis 1. As is evident in the character trait list, there are multiple head shapes and sizes, arm and leg positions and proportions, breast and buttock shapes, as well as decoration types and locations. Additionally, this hypothesis predicts a high amount of homoplasy and that statuettes should group indiscriminately on the tree. If this hypothesis were correct, the RI should be much closer to zero. An RI of .443 indicates that some branching is occurring. Also, the fact that the Venus of Pavlov, the Lozenge, Moravany, the Bicephalous, and Lespugue form terminal branches demonstrates how unique each of these figurines is.

While the Italian figurines (Savignano, the Punchinello, and the Abrachial) have grouped together the way that hypothesis 1 would suggest, these statuettes are grouped together on the basis of a few generalized characters (analogous to biological primitive characters). Because there are only so many ways to create an anthropomorphic object, the grouping of general features might not suggest communication between groups in this case. Instead, certain figurines may be grouping together due to what I would call stylistic convergence (see Fig. 5), where two unrelated groups

**Fig. 5** Stylistic convergence (While the statuettes might appear morphologically similar, they are from two separate archaeological cultures. The statuette on the left is from a site called Avaritsa in Greece and is Neolithic, No GR157b 1909, material not specified, ~120 cm (Ucko 1968: 493). The marl figurine on the right is Gravettian and from New Avdeevo, Russia No 11, ~58 cm (Gvozdover 1995: 145))

are trying to represent a similar idea and produce similar imagery due to various limitations/constraints. These constraints could be due to the material utilized, skill of the maker, similar cultural reasons (i.e., to represent pregnancy), or whether or not the figurine was finished. For example, Fig. 5 depicts two statuettes that are very generalized and are also lacking decorations. While they may look very similar, only one figurine was created in the Gravettian, indicating the independent innovation of similar images. I believe this is the reason that the Punchinello and Avdeevo 78–9 are sister groups as well as Khotylevo 2–1 and the Abrachial, as a result of cultural homoplasy. All of these statuettes are full-figured women with large breasts, undecorated bodies, broad hips, and protruding abdomens. It is also possible that if all the Italian statuettes were utilized, instead of just three, they may have grouped together. However, I was only able to include figurines that were complete and those that I had access to viewing from the front, back, and profiles.

Hypothesis 2, by Gvozdover (1989), also suggests some blending among cultures – but specifically those of the Russian Plains. Hypothesis 2 is supported by the data as the Russian statuettes predominately group together to the exclusion of statuettes from other regions. These statuettes have clustered based on a unique combination of specialized characters (analogous to biological-derived characters). For example, Zaraysk, Avdeevo 77–1, Avdeevo 76–6, and Avdeevo 48–3 group together. They are characterized by sharing the following features: enlarged head, the same head shape frontal and profile, completeness of arms, position of arms, protruding abdomen, broad hips, feet that are turned inward, legs that are apart below the knees, complete legs (with feet), and miniature legs. In this example, the statuettes share similar morphological features and are similar in proportion throughout the body. Another Russian grouping includes Kostenki 83–1, Khotylevo 22, and Kostenki 4. Theses statuettes all have large heads that share the same shape in frontal and profile views, heads that face downward, lack facial features, share breast shape, arms that lack hands, and long arms. Both of these groups are significant because they suggest that specific art-making methods or traditions were in place when creating figurines. The combination of features also seems culturally significant and may suggest that these particular traits are related to the meaning of these statuettes.

Lastly, it is important to note that both hypotheses predicted blending among Gravettian cultures and did not predict strict cultural phylogenesis at the site level. The cladogram corroborates this point, as archaeological sites do not form monophyletic groups. Instead, the results of the cladogram demonstrate a blending of the Eastern European sites.

## 6 Conclusions and Areas of Future Research

While we may never know the meaning that the makers of the Venuses were trying to convey, it is obvious that these populations intentionally copied particular details when making them. In discussing cultural evolution, Shennan (2008) suggested why particular aspects are copied and reproduced among one population. He argued that it could be due to a prestige bias, in which individuals copy a person that is important. Or, he argued, people reproduce cultural behaviors and ideals due to a conformist bias or because that is what is practiced locally.

The archaeological evidence supports a connection between sites in the Russian Plains. Many authors have discussed what is known as the Kostenki-Avdeevo culture (Soffer 1985; Grigor'ev 1993; Soffer and Praslov 1993; Iakovleva 2000). Gravettian figurative art in the Russian Plains comes from two different regions: the Desna and the Don. Avdeevo I and II and Khotylevo II belong to the Desna group and Kostenki I layer one, Zaraysk, and Gagarino to the Don (Iakovleva 2000; Amirkhanov 2008). The majority of these sites have rich artistic inventories, which frequently include decorated ivory plaques, animal statuettes, female statuettes, and various decorated objects. Female figurines are generally found in multiples at these sites. These assemblages also share similarities in their layouts; many contain the inferred remains of dwellings with multiple hearths and pits. These sites have similar lithic inventories and material cultures as well. In fact, three scrapers made of an exotic raw material located near Kostenki (nearly 120.70 km downstream) were also found at Gagarino (author unknown 1942). Additionally, statuettes from these sites were all found in habitation structures, thus providing some contextual detail (Iakovleva 2000). All of the aforementioned data suggest that individuals at these sites were interacting with each other and shared a cultural unity. Although art making in the Western world is often associated with individuality, in some societies art is used to express a collective image, which contributes to a group identity (Phillips and Steiner 1999). Thus, it makes sense that we do not see separate branches for each site (Gagarino, Kostenki, Khotylevo II, and Avdeevo).

In contrast to this last point, several of the statuettes (Venus of Pavlov, the Lozenge, and the Bicephalous) diverge and do not have a single sister group. It seems that these statuettes have many unique features. In the case of the Venus of Pavlov, it is either extremely abstract or unfinished. Both the Lozenge and the Bicephalous come from the same site, Balzi Rossi (Italy). Unlike the figurines in many of the other Eurasian collections, the majority of the Balzi Rossi statuettes were intentionally unrealistic representations of females. The Lozenge has a diamond shaped body with triangular legs, while the Bicephalous is a double figurine with two heads and exaggerated sexual features. It is possible that these particular statuettes were not tied to group identity but may have represented a particular character (either mythical or earthly). It seems that these figurines were created in cultures where individuals more freely expressed alternative morphological features.

The Balzi Rossi figurines may however still provide us with reflections about group identity. Porr (2010: 150) has argued "that it is crucial to see the statuettes in relation to their actual use and consequently their relationships to bodily practices and corporal culture." Therefore, it is not enough to only look for morphological similarities on the figurines; we must also look for functional clues to provide meaning and understand their cultural relevance. The majority of the figurines appear to be pendants. In fact, of the 15 pieces from Grimaldi, Mussi and colleagues (2000) noted that six have clear perforations, and seven others possibly had them, but because of the incompleteness at the distal ends, it is impossible to say for sure.

Evidence of polish is present on the two double statuettes, the Doublet and the Bicephalous, suggesting use-wear. In both cases, the polish is found in the holes between the two heads of the specimens. This indicates that the objects could have been either suspended from a string, worn on the body, or attached to an object.

Too many hypotheses surrounding the "Venus" figurines have begun with the assumption that the figurines share core similarities but offer little explanation as to what those features are. This analysis has shown that in fact there are regional connections in art making among sites within the Russian Plains but also documents the uniqueness and diversity of individual statuettes from other regions. Importantly, these data also demonstrate that the figurines are not "practically interchangeable" as argued by Leroi-Gourhan. In order to make larger connections about what these figurines were used for, we need to start by analyzing individual statuettes. After the individual is analyzed in morphology and function, the researcher should then look at the figurines at a site level before looking at the regional or continental levels. Many of these figurines are found at sites that also possess animal statuettes. Perhaps by overemphasizing a shared meaning among these anthropomorphic objects, to the detriment of investigating animal representations and possible associations between the human and animal figures, we are creating a dichotomy that never existed in the Paleolithic. In terms of homogeneity and diversity in the Gravettian imagery, the patterns in the data suggest that the issue is far more complex than many have hypothesized.

While I feel that the character trait list is substantial, it is also possible to add more traits, in order to increase resolution. For example, the following traits could be added: presence or absence of a neck, chin, type of hair, and type of decoration on arms and legs. I was not able to gain access to study the entire Gravettian collection. Future studies could include more complete figurines from Italy and Central Europe. Adding more specimens could give new insights into unique shared features that could not be seen in this cladogram.

Since it has been demonstrated that there was an art-making tradition that extended within the Russian Plains, it would be interesting to compare statuettes from the Russian Plains with those from Mal'ta. The statuettes from Mal'ta are a bit younger than the rest of the Gravettian figurines and have been argued to be stylistically different (Leroi-Gourhan 1968). The Mal'ta figurines are interpreted as being fully clothed, having facial features, and being probably worn as pendants hung upside down. If the two regions formed separate groups, it could quantitatively validate these arguments. It may also be beneficial to compare Gravettian and Magdalenian figurines to determine which, if any, features distinguish statuettes from each culture. Or a different character list could be created that allows for the analysis of both animal and human statuettes, since both are often found together at various sites.

Lastly, cladistics approaches could also be applied to figurines from other archaeological cultures or other forms of Paleolithic art. For example, a variety of ceramic statuettes were created by the Jomon culture of the Japanese archipelagos. This culture dates to 16,500 years ago and immediately follows the Paleolithic (Habu 2004). The Jomon people were part of large settlements and can be described as hunter-gatherer-fisherman. Figurines were created in the Early, Middle, and Late Jomon cultures and are assumed to have religious significance (Maringer 1974). They vary in appearance depending on when they were created, with earlier figurines being more theriomorphic and later figurines representing females (Maringer 1974). Since the Jomon were involved in extensive trade networks, it would be interesting to use cladistics to analyze how different features have changed over time and dispersed across various settlements.

**Acknowledgments** This research could not have been completed without a grant from Leiden University Fund (LUF). I would also like to thank Dr. Wil Roebroeks for his support and for helping me get into contact with the curators at Monrepos, the Kunstkamera, and the Hermitage. I am also grateful to Dr. Kathy MacDonald's constructive criticism. I would also like to thank everyone that helped me in Saint Petersburg including Svetlana Demeshchenko and Dr. Gennady Khlopachev at the Kunstkamera. Thank you to Dr. Sergey Vasil'ev for arranging my visa invitation and permission to do research at both museums in Russia. I would also like to thank Olaf Jöris and Sabine Gaudzinkski for their help at Monrepos. Thank you to Thayne Werdel for helping me with TNT; I never would have figured it out without your help. Lastly, thank you to Rutger Jansma for the inspiration for the project.

# 7 Appendix 1: Character List

| | |
|---|---|
| *Head* | |
| 1. Facial features | (0) absent, (1) present |
| 2. Orientation | (0) forward, (1) downward |
| 3. Proportion | (0) miniature, (1) normal, (2) enlarged |
| 4. Head shape (profile) | (0) round, (1) intermediate, (2) club, (3) pointed, (4) flat |
| 5. Head shape (frontal) | (0) round, (1) intermediate, (2) club, (3) pointed, (4) flat |
| *Torso* | |
| 6. Proportion | (0) miniature, (1) normal, (2) enlarged |
| 7. Belly button | (0) absent, (1) present |
| *Arms and hands* | |
| 8. Completeness | (0) complete, (1) hand absent, (2) hand + forearm absent |
| 9. Proportion | (0) miniature, (1) normal, (2) enlarged |
| 10. Arm position | (0) absent, (1) at sides, (2) above breasts, (3) below breasts, (4) on abdomen, (5) on hips, (6) away from body, |
| 11. Fingers | (0) absent, (1) present |
| *Legs and feet* | |
| 12. Completeness | (0) complete, (1) feet absent, (2) feet + lower leg absent |
| 13. Apart | (0) absent, (1) ^ knees, (2) ˇ knees, (3) at knees |
| 14. Proportion | (0) miniature, (1) normal, (2) enlarged |
| 15. Realism | (0) normal, (1) abstract |
| 16. Knees turn inward | (0) absent, (1) present |
| 17. Puncture in place of feet | (0) absent, (1) present |
| 18. Toes | (0) absent, (1) present |

| | |
|---|---|
| *Sexual features* | |
| 19. Breasts | (0) elongated, (1) intermediate, (2) round |
| 20. Belly | (0) absent, (1) pronounced |
| 21. Hips | (0) narrow, (1) broad |
| 22. Pubic triangle | (0) absent, (1) present |
| 23. Vulva | (0) absent, (1) present |
| 24. Buttock shape (profile) | (0) absent, (1) flat, (2) shelf, (3) intermediate, (4) round |
| 25. Buttock shape (rear) | (0) absent, (1) flat, (2) heart, (3) intermediate, (4) round |
| *Decorations* | |
| 26. Head (type) | (0) absent, (1) incision, (2) hat, (3) puncture, (4) hair |
| 27. Torso (front) (type) | (0) absent, (1) incision, (2) rope, (3) puncture |
| 28. Torso (back) (type) | (0) absent, (1) incision, (2) rope, (3) puncture |
| 29. Arms | (0) absent, (1) present |
| 30. Legs | (0) absent, (1) present |

# 8 Appendix 2: Character States That Are Less Straightforward

Head Shape Profile

0) Absent 1) Round 2) Club 3) Intermed 4)Pointed

5) Flat

Head Shape (Frontal)

0) Absent 1) Round 2) Intermed 3) Pointed 4) Square

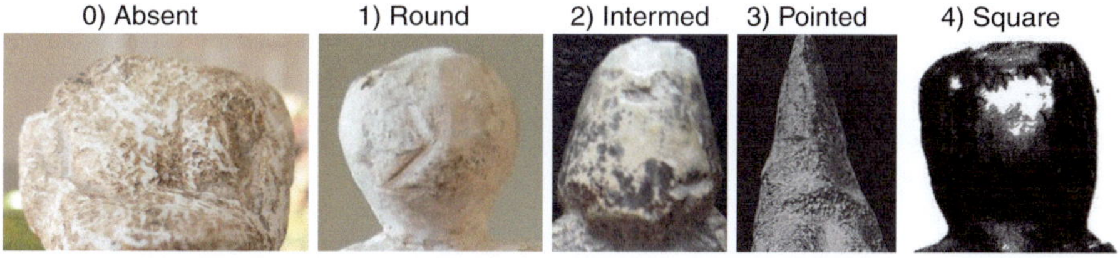

Completeness of Arms

0) Absent 1) Complete 2) Hand absent 3) Hand + forearm absent

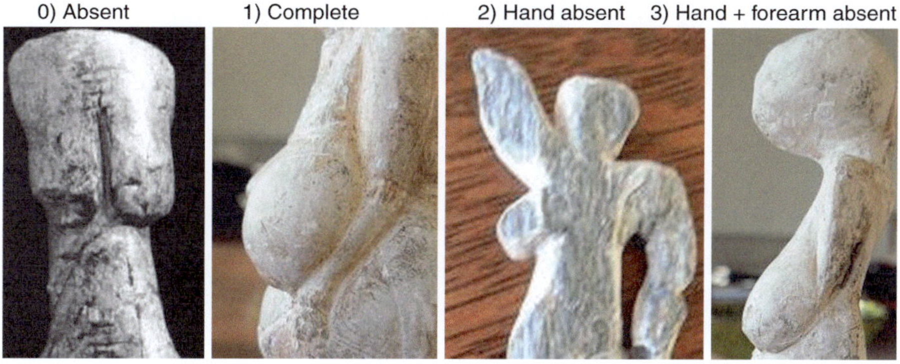

Arm Position

0) Absent

1) At sides

2) Above breasts

3) Below breasts

4) On abdomen

5) On hip

6) Away from Body

Completeness of Legs

0) Absent

1) Complete

2) Without feet

3) Feet + lower leg absent

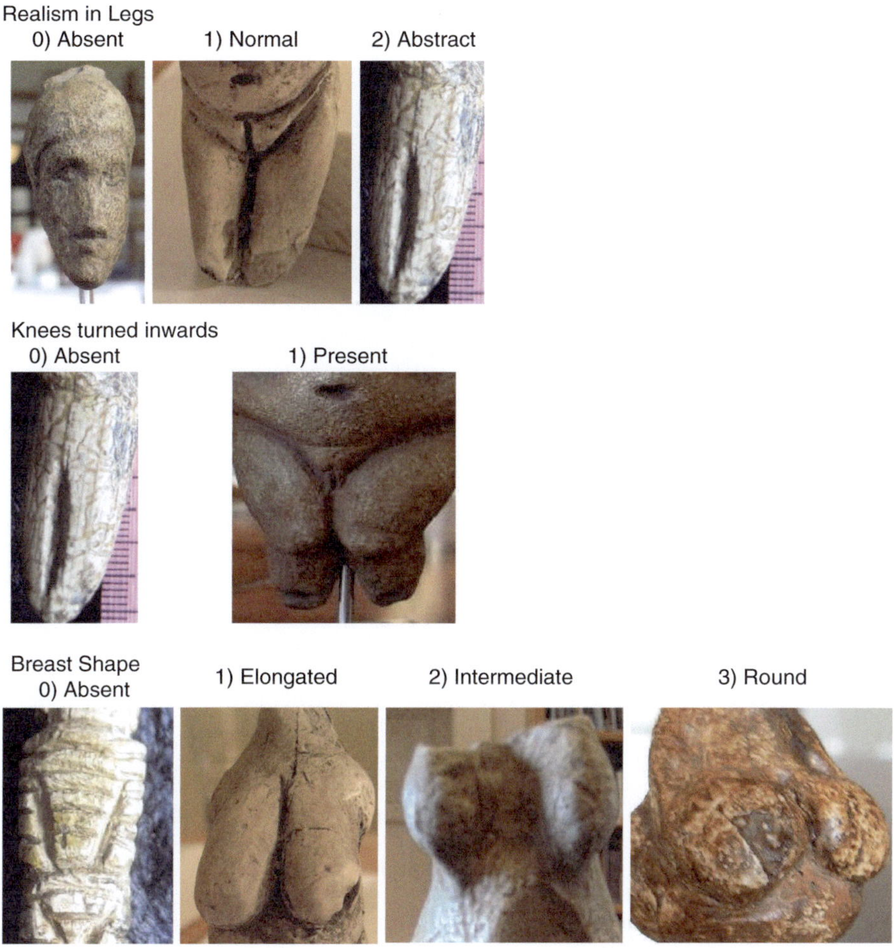
Realism in Legs
0) Absent
1) Normal
2) Abstract
Knees turned inwards
0) Absent
1) Present
Breast Shape
0) Absent
1) Elongated
2) Intermediate
3) Round

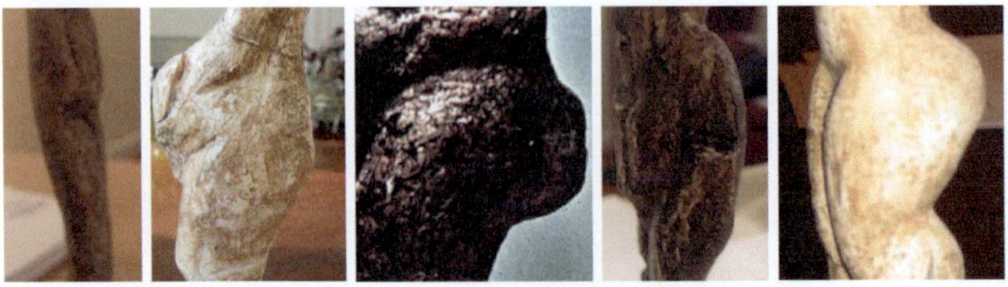


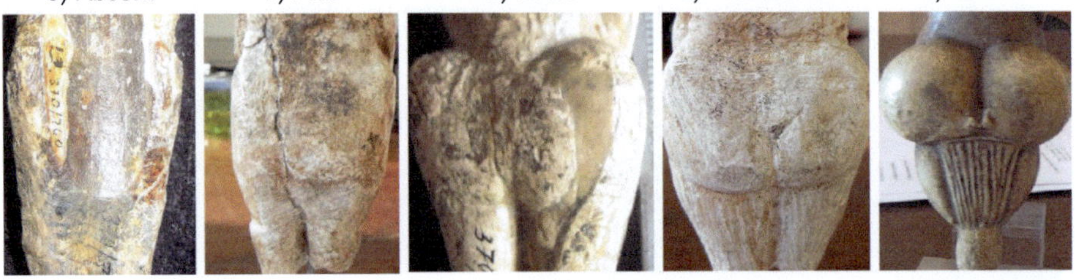


## References

Ambramova, Z. (1995). *L'art Paléolithique D'europe Orientale et de Sibérie. Collection L'Homme des Origines*. Grenoble: Jérôme Millon.

Amirkhanov, H. and S. Lev (2008). A unique Paleolithic sculpture from the site of Zaraysk (Russia). *Antiquity*, 76, 613–614.

Archie, J. W. (1989). A randomization test for phylogenetic information in systematic data. *Systematic Zoology*, *38*, 219–252.

Author unknown. (1942). Prehistory in the U.S.S.R. I. Paleolithic and mesolithic. B. the Russian Plain. *Man*, *60*, 100–103.

Bamforth, D. B. (2002). Evidence and metaphor in evolutionary archaeology. *American Antiquity*, *67*, 435–452

Bégouen, Count. (1925). The magic origin of prehistoric art. *Antiquity*, *3*, 5–19.

Buchanan, B., & Collard, M. (2007). Investigating the peopling of North America through cladistic analyses of Early Paleo-Indian projectile points. *Journal of Anthropological Archaeology*, *26*, 366–393.

Buss, D (2004). Evolutionary Psychology the New Science of the Mind. Boston: Pearson Education Inc.

Campbell, N., J.B. Reece, E.J. Simon, and J.L. Dickey (2003). Biology Concepts and Connections. Fourth edition. San Francisco: Pearson Education Inc.

Collard, M., & Shennan, S. (2000). Processes of culture change in prehistory: A case study from the Europe Neolithic. In C. Renfrew, K. Boyle (Eds.), *Archaeogenetics: DNA and the population prehistory of Europe* (pp. 89–97). Cambridge: McDonald Institute for Archaeological Research.

Collard, M., Shennan, S. J., & Tehrani, J. (2006). Branching verses blending in macroscale cultural evolution: A comparative study. In C. P. Lipo, M. J. O'Brien, M. Collard, S. J. Shennan (Eds.),

*Phylogenetic approaches in anthropology and prehistory mapping our ancestors* (pp. 53–64). New Brunswick: Aldine Transaction.
Conard, N. (2009). A female figurine from the basal Aurignacian of Hohle Fels Cave in southwestern Germany. *Nature*, *459*, 248–252.
Delporte, H. (1993). *L'image de la femme dans l'arte préhistorique*. Paris: Picard.
Duhard, J. (1993). Upper paleolithic figurines as a reflection of human morphology and social organization. *Antiquity*, *67*, 83–91.
Eerkens, J. W., Bettinger, R. L., & McElreath, R. (2006). Cultural transmission, phylogenetics, and the archaeological record. In C. P. Lipo, M. J. O'Brien, M. Collard, & S. J. Shennan (Eds.), *Phylogenetic approaches in anthropology and prehistory mapping our ancestors* (pp. 169–183). New Brunswick: Aldine Transaction.
Farris, J. (1989). The retention index and the rescaled consistency index. *Cladistics*, *5*, 417–419.
Fracchia, J., & Lewontin, R. C. (1999). Does culture evolve? *History and Theory Theme, 38*, 52–78.
Fracchia, J., & Lewontin, R. C. (2005). The price of metaphor. *History and Theory*, *44*, 14–29.
Gamble, C. (1982). Interaction and alliance in paleolithic society. *Man*, *17*, 92–107.
Goloboff, P., Farris, S., & Nixon, K. (2000). *TNT (Tree Analysis using New Technology) (BETA) ver. Xxx Published by the authors*. Tucumán, Argentina.
Goloboff, P., Farris, J., & Nixon K.C. (2008). T.N.T. Free program for phylogenetic analysis. *Cladistics*, 24(5):774–786.
Grigor'ev, G. P. (1993). *The Kostenki-Avdeevo archaeological cultural unity. From Kostenki to Clovis*. New York: Plenum Press.
Guthrie, R. D. (2005). *The nature of Paleolithic art*. Chicago: University of Chicago Press.
Gvozdover, M. (1989). The typology of figurines from the Kostenki Paleolithic. *Anthropology and Archeology of Eurasia*, *27*(4), 32–94.
Gvozdover, M. (1995). Art of the Mammoth Hunters the Finds from Avdeevo. Oxbow monograph 49. Oxford: Oxbox Books.
Habu, J. (2004). *The Ancient Jomon of Japan* (p. 332) Cambridge: Cambridge University Press.
Hahn, J. (1990). Fonction et signification des statuettes du Paléolithique superior. In J.Clottes (eds.), *L'art des objects au Paléolithique*, t.2, 173–183, Paris: Direction du Patrimoine.
Hawkes, J., & Wolley, L. (1963). History of Mankind cultural and scientific development. In *Prehistory and the beginnings of civilization* (Vol. 1). New York: Harper and Row.
Holden, C. J., & Shennan, S. (2005). Introduction to Part 1: How tree-like is cultural evolution? In H. Mace, & S. Shennan (Eds.), *The evolution of cultural diversity a phylogenetic approach*. Walnut Creek: Left Coast Press.
Hudson, D. H., & Scornavacca, C. (2012). Dendroscope 3: An interactive tool for rooted phylogenetic trees and networks. *Systematic Biology*, *61*(6):1061–1067.
Iakovleva, L. (2000). The Gravettian art of Eastern Europe as exemplified in the figurative art of Kostenki I. In *Hunters of the Golden Age*. Leiden: Leiden University, Eds. Wil Roebroeks, Margherita Mussi, Jiri Svoboda, & Kelly Fennema.
Jordan, P. (2009). Linking pattern to process in cultural evolution: Investigating material culture diversity among the Northern Khanty of Northwest Siberia. In S. Shennan (Ed.), *Pattern and process in cultural evolution*. Berkeley: University of California Press.
Jordan, P., & Shennan, S. (2003). Cultural transmission, language, and basketry traditions amongst the California Indians. *Journal of Anthropological Archaeology*, *22*, 42–74.
Koenigswald, G. H. R. von (1972). Early Homo sapiens as an artist: The meaning of Paleolithic art. In *The origin of homo sapiens, ecology and conservation* (Vol. 3). Proceedings of the Paris Symposium 1969. France, F. Border, ed. pp. 133–139.
Leroi-Gourhan, A. (1968). *The art of prehistoric man in western Europe*. London: Thames & Hudson.
Leroi-Gourhan, A. (1982). The Dawn of European Art: An Introduction to Palaeolithic Cave Painting. Cambridge: Cambridge University Press.
Levy, G. (1948). *The Gate of the Horn: A study of the religious conceptions of the stone age, and their influence upon European thought*. London: Faber & Faber.

Maringer, J. (1974). Clay figurines of the Jōmon period: A contribution to the history of ancient religion in Japan. *History of Religions*, *14*(2), 128–132.
Markale, Jean. (1999) The Great Goddess: Reverence of the Divine Feminine from the Paleolithic to the Present. Vermont: Inner Traditions International.
McCoid, C., & McDermott, L. (1996). Towards decolonizing gender: Female vision in the upper paleolithic. *American Anthropologist*, *New Series*, *98*(2): 319–326
McDermott, L. (1996). Self-representation in upper paleolithic female figurines. *Current Anthropology*, *37*, 227–275.
Mussi, M., et al. (2000). Echoes from the Mammoth Steppe. In: *Hunters of the golden age*. Leiden: Leiden University.
Nelson, S. (1990). Diversity of the upper paleolithic "Venus" figurines and archaeological mythology. In C. B. Brettel, & C. F. Sargent (Eds.), *Gender in cross-cultural perspective* (pp. 51–58). Englewood Cliffs: Prentice-Hall.
Nunn, C. L., Arnold, C., Matthews, L., Mulder, M. B. (2010). Simulating trait evolution for cross-cultural comparison. *Philosophical Transactions of the Royal Society B: Biological Sciences*, *365*(1559), 3807–3819.
O'Brien, M., & Lyman, R. (2003). *Cladistics and archaeology*. Salt Lake City: The University of Utah Press.
O'Brien, M. J., Darwent, J., Lyman, R. L. (2001). Cladistics is useful for reconstructing archaeological phylogenies: Paleoindian points from the Southwest United States. *Journal of Archaeological Science*, *28*, 1115–1136.
Pettitt, P. (2000). Chronology of the mid upper paleolithic: The radiocarbon evidence. In *Hunters of the golden age*. Leiden: Leiden University.
Phillips, R. B., & Steiner, C. B. (1999). *Unpacking culture: Art and commodity in colonial and postcolonial worlds*. Berkeley: University of California Press.
Porr, M. (2010). The Hohle Fels "venus": Some remarks on animals, humans and metaphorical relationships in early upper Paleolithic art. *Rock Art Research*, *27*(2), 147–159.
Reinach, S. (1903). L'art et la magie. A propos des peintures et des gravures de l'Age du Renne. *L'Anthropologie*, *14*, 257–66.
Rexova, K. Frynta, D., Zrzavy, J. (2003). Cladistic analysis of languages: Indo-European classification based on lexicostatistical data. *Cladistics*, 19(2), 120–127.
Rice, P. (1981). Prehistoric venuses, symbols of motherhood or womanhood? *Journal of Anthropological Research*, *37*(4), 402–414.
Shennan, S. (2008). Evolution in archaeology. *Annual Review of Anthropology*, *37*, 75–91.
Soffer, O. (1985). *The upper Paleolithic of the central Russian Plain*. San Diego: Academic Press Inc.
Soffer, O. (1987). Upper Paleolithic Connubia, Refugia, and the Archaeological Record of Eastern Europe. In *The Pleistocene Old World Interdisciplinary Contributions to Archaeology*. Springer: U.S. p.333–348.
Soffer, O., & Praslov, N. D. (1993). *From Kostenki to Clovis*. New York: Plenum Press.
Svoboda, J., Klíma, B., Jaroŝova, L., & Ŝkrdla, P. (2000). The Gravettian in Moravia: Climate, Behavior, and Technological Complexity. In *Hunters of the Golden Age*. Leiden: Leiden University. Eds. Wil Roebroeks, Margherita Mussi, Jiři Svoboda.
Taylor, T. (1996). The Prehistory of Sex: Four Million Years of Human Culture. London: Fourth Estate.
Tehrani, J., & Collard, M. (2002). Investigating cultural evolution through biological phylogenetic analyses of Turkman textiles. *Journal of Anthropological Archaeology*, *21*, 443–463.
Tripp, A. J., & Schmidt, N. (2013). Analyzing fertility and attraction in the paleolithic: The venus figurines. *Archaeology, Ethnology and Anthropology of Eurasia, 41*(1), 54–60.
Ucko, P. (1962). The interpretation of prehistoric anthropomorphic figurines. *Journal of the Royal Institute of Great Britain and Ireland*, *92*, 38–54.
Ucko, Peter J (1968). Anthropomorphic Figurines. London: Andrew Szmidla.
Verpoorte, A. (2001). *Places of art, traces of fire a contextual approach to anthropomorphic figurines in the Pavlovian (Central Europe, 29–24 kyr BP)*. Leiden: Leiden University Press.
White, R. (2006). The women of brassempouy: A century of research and interpretation. *Journal of Archaeological Method and Theory*, *13*(4), 251–304.

MIX
Papier aus verantwortungsvollen Quellen
Paper from responsible sources
FSC® C105338

If you have any concerns about our products,
you can contact us on
**ProductSafety@springernature.com**

In case Publisher is established outside the EU,
the EU authorized representative is:
**Springer Nature Customer Service Center GmbH**
**Europaplatz 3, 69115 Heidelberg, Germany**

Printed by Libri Plureos GmbH
in Hamburg, Germany